Communications
in Computer and Information Science 1427

More information about this series at http://www.springer.com/series/7899

Alexander Sychev · Sergey Makhortov ·
Bernhard Thalheim (Eds.)

Data Analytics and Management in Data Intensive Domains

22nd International Conference, DAMDID/RCDL 2020
Voronezh, Russia, October 13–16, 2020
Selected Proceedings

 Springer

Editors
Alexander Sychev
Voronezh State University
Voronezh, Russia

Sergey Makhortov
Voronezh State University
Voronezh, Russia

Bernhard Thalheim (iD)
Christian-Albrecht University of Kiel
Kiel, Schleswig-Holstein, Germany

ISSN 1865-0929   ISSN 1865-0937 (electronic)
Communications in Computer and Information Science
ISBN 978-3-030-81199-0   ISBN 978-3-030-81200-3 (eBook)
https://doi.org/10.1007/978-3-030-81200-3

This Springer imprint is published by the registered company Springer Nature Switzerland AG
The registered company address is: Gewerbestrasse 11, 6330 Cham, Switzerland

Preface

This CCIS volume published by Springer contains the proceedings of the XXII International Conference on Data Analytics and Management in Data Intensive Domains (DAMDID/RCDL 2020) that was set to be held at the Voronezh State University, Russia, during October 13–16, 2020. However, because of the worldwide COVID-19 crisis, DAMDID/RCDL 2020 had to take place online.

DAMDID is a multidisciplinary forum of researchers and practitioners from various domains of science and research, promoting cooperation and the exchange of ideas in the area of data analysis and management in domains driven by data-intensive research. Approaches to data analysis and management being developed in specific data-intensive domains (DID) of X-informatics (such as X = astro, bio, chemo, geo, med, neuro, physics, chemistry, material science, etc.) and social sciences, as well as in various branches of informatics, industry, new technologies, finance, and business, contribute to the conference content.

Previous DAMDID/RCDL conferences were held in St. Petersburg (1999, 2003), Protvino (2000), Petrozavodsk (2001, 2009), Dubna (2002, 2008, 2014), Pushchino (2004), Yaroslavl (2005, 2013), Pereslavl (2007, 2012), Kazan (2010, 2019), Voronezh (2011), Obninsk (2016), and Moscow (2017, 2018).

The program of DAMDID/RCDL 2020 was oriented towards data science and data-intensive analytics as well as data management topics. The proceedings include abstracts of the two keynotes.

The keynote by Ladjel Bellatreche (full professor at the National Engineering School for Mechanics and Aerotechnics, France) was aimed to motivate researchers to embrace the energy-efficiency of data management systems. Oscar Pastor (full professor and director of the Research Center on Software Production Methods at the Polytechnic University of Valencia, Spain) gave a talk discussing a fundamental role of a sound conceptual modeling in human genome research.

The conference Program Committee, comprised of members from 8 countries, reviewed 60 submissions. 29 submissions were accepted as full papers, 14 as short papers, and 3 as short demos, whilst 14 submissions were rejected.

According to the conference and workshops program, 55 oral presentations were grouped into 11 sessions. Most of the presentations were dedicated to the results of research conducted in Russian organizations including those located in Dubna, Ekaterinburg, Innopolis, Kazan, Krasnodar, Moscow, Novosibirsk, Obninsk, Tomsk, Voronezh, Pereslavl, Perm, St. Petersburg, Petrozavodsk, Tver, Tyumen, and Yaroslavl. However, the conference also featured talks prepared by the foreign researchers from countries such as Armenia, France, Germany, Spain, UK, and USA.

For the CCIS conference proceedings, 16 peer reviewed papers and 2 keynote abstracts have been selected by the Program Committee (acceptance rate: 27%) and-structured into 6 sections including Data Integration, Conceptual Models, and Ontologies (3 papers); Data Management in Semantic Web (3 papers); Data Analysis in

Medicine (2 papers); Data Analysis in Astronomy (3 papers); and Information Extraction from Text (5 papers).

We are grateful to the Program Committee members, for reviewing the submissions and selecting the papers for presentation, to the authors of the submissions, and to the host organizers from Voronezh State University. We are also grateful for the use of the Conference Management Toolkit (CMT) sponsored by Microsoft Research, which provided great support during various phases of the paper submission and reviewing process.

April 2021   Alexander Sychev
                                                               Sergey Makhortov
                                                              Bernhard Thalheim

Organization

Program Committee Co-chairs

Sergey Makhortov	Voronezh State University, Russia
Alexander Sychev	Voronezh State University, Russia
Bernhard Thalheim	University of Kiel, Germany

Program Committee Deputy Chair

Sergey Stupnikov — Federal Research Center "Computer Science and Control" of RAS, Russia

PhD Workshop Chair

Sergey Makhortov — Voronezh State University, Russia

PhD Workshop Curator

Ivan Lukovic — University of Novi Sad, Serbia

Organizing Committee Co-chairs

Oleg Kozaderov — Voronezh State University, Russia
Victor Zakharov — Federal Research Center "Computer Science and Control" of RAS, Russia

Organizing Committee Deputy Chairs

Eduard Algazinov (Deceased) — Voronezh State University, Russia
Sergey Stupnikov — Federal Research Center "Computer Science and Control" of RAS, Russia

Organizing Committee

Andrey Koval — Voronezh State University, Russia
Nikolay Skvortsov — Federal Research Center "Computer Science and Control" of RAS, Russia
Dmitry Borisov — Voronezh State University, Russia
Alexey Vakhtin — Voronezh State University, Russia
Dmitry Briukhov — Federal Research Center "Computer Science and Control" of RAS, Russia
Olga Schepkina — Voronezh State University, Russia

Coordinating Committee

Igor Sokolov (Co-chair)	Federal Research Center "Computer Science and Control" of RAS, Russia
Nikolay Kolchanov (Co-chair)	Institute of Cytology and Genetics, SB RAS, Russia
Sergey Stupnikov (Deputy Chair)	Federal Research Center "Computer Science and Control" of RAS, Russia
Arkady Avramenko	Pushchino Radio Astronomy Observatory, RAS, Russia
Pavel Braslavsky	Ural Federal University, SKB Kontur, Russia
Vasily Bunakov	Science and Technology Facilities Council, UK
Alexander Elizarov	Kazan Federal University, Russia
Alexander Fazliev	Institute of Atmospheric Optics, SB RAS, Russia
Alexei Klimentov	Brookhaven National Laboratory, USA
Mikhail Kogalovsky	Market Economy Institute, RAS, Russia
Vladimir Korenkov	Joint Institute for Nuclear Research, Russia
Mikhail Kuzminski	Institute of Organic Chemistry, RAS, Russia
Sergey Kuznetsov	Institute for System Programming, RAS, Russia
Vladimir Litvine	Evogh Inc., USA
Archil Maysuradze	Moscow State University, Russia
Oleg Malkov	Institute of Astronomy, RAS, Russia
Alexander Marchuk	Institute of Informatics Systems, SB RAS, Russia
Igor Nekrestjanov	Verizon Corporation, USA
Boris Novikov	Saint Petersburg State University, Russia
Nikolay Podkolodny	Institute of Cytology and Genetics, SB RAS, Russia
Aleksey Pozanenko	Space Research Institute, RAS, Russia
Vladimir Serebryakov	Computing Center of RAS, Russia
Yury Smetanin	Russian Foundation for Basic Research, Moscow
Vladimir Smirnov	Yaroslavl State University, Russia
Konstantin Vorontsov	Moscow State University, Russia
Viacheslav Wolfengagen	National Research Nuclear University "MEPhI", Russia
Victor Zakharov	Federal Research Center "Computer Science and Control" of RAS, Russia

Program Committee

Ladjel Bellatreche	National Engineering School for Mechanics and Aerotechnics, France
Dmitry Borisenkov	Relex, Russia
Pavel Braslavski	Ural Federal University, Russia
Vasily Bunakov	Science and Technology Facilities Council, UK
George Chernishev	Saint Petersburg State University, Russia
Boris Dobrov	Lomonosov Moscow State University, Russia
Alexander Elizarov	Kazan Federal University, Russia

Alexander Fazliev	Institute of Atmospheric Optics, SB RAS, Russia
Yuriy Gapanyuk	Bauman Moscow State Technical University, Russia
Veronika Garshina	Voronezh State University, Russia
Evgeny Gordov	Institute of Monitoring of Climatic and Ecological Systems, SB RAS, Russia
Valeriya Gribova	Far Eastern Federal University, Russia
Maxim Gubin	Google Inc., USA
Sergio Ilarri	University of Zaragoza, Spain
Mirjana Ivanovic	University of Novi Sad, Serbia
Nadezhda Kiselyova	IMET RAS, Russia
Vladimir Korenkov	Joint Institute for Nuclear Research, Russia
Sergey Kuznetsov	Institute for System Programming, RAS, Russia
Evgeny Lipachev	Kazan Federal University, Russia
Natalia Loukachevitch	Lomonosov Moscow State University, Russia
Ivan Lukovic	University of Novi Sad, Serbia
Oleg Malkov	Institute of Astronomy, RAS, Russia
Yannis Manolopoulos	Aristotle University of Thessaloniki, Greece
Archil Maysuradze	Lomonosov Moscow State University, Russia
Manuel Mazzara	Innopolis University, Russia
Alexey Mitsyuk	National Research University Higher School of Economics, Russia
Xenia Naidenova	Kirov Military Medical Academy, Russia
Dmitry Namiot	Lomonosov Moscow State University, Russia
Dmitry Nikitenko	Lomonosov Moscow State University, Russia
Panos Pardalos	University of Florida, USA
Natalya Ponomareva	Research Center of Neurology, Russia
Alexey Pozanenko	Space Research Institute, RAS, Russia
Roman Samarev	Bauman Moscow State Technical University, Russia
Timos Sellis	Swinburne University of Technology, Australia
Vladimir Serebryakov	Computing Centre of RAS, Russia
Nikolay Skvortsov	Federal Research Center "Computer Science and Control" of RAS, Russia
Manfred Sneps-Sneppe	AbavaNet, Russia
Sergey Sobolev	Lomonosov Moscow State University, Russia
Valery Sokolov	Yaroslavl State University, Russia
Alexey Ushakov	University of California, Santa Barbara, USA
Pavel Velikhov	Huawei, Russia
Alexey Vovchenko	Federal Research Center "Computer Science and Control" of RAS, Russia
Vladimir Zadorozhny	University of Pittsburgh, USA
Yury Zagorulko	Institute of Informatics Systems, SB RAS, Russia
Victor Zakharov	Federal Research Center "Computer Science and Control" of RAS, Russia
Sergey Znamensky	Institute of Program Systems, RAS, Russia
Mikhail Zymbler	South Ural State University, Russia

Supporters

Voronezh State University
Federal Research Center "Computer Science and Control" of the Russian Academy of Sciences (FRC CSC RAS), Moscow, Russia
Moscow ACM SIGMOD Chapter

Keynote Speakers' Bios

Ladjel Bellatreche (National Engineering School for Mechanics and Aerotechnics of Aerospace Engineering Group, France)

Ladjel Bellatreche is a Full Professor at National Engineering School for Mechanics and Aerotechnics (ENSMA) of Aerospace Engineering Group (ISAE), Poitiers, France, where he joined as a Faculty Member since September 2010. He leads the Data Engineering Group of the Laboratory of Computer Science & Automatic Control for Systems (LIAS). He was Director of the LISI laboratory from January 2011 till January 2012. Prior to that, he spent eight years as an Assistant and then an Associate Professor at the Poitiers University, France. Ladjel was a Visiting Professor at the Harbin Institute of Technology (HIT), China (2018, 2019), the Québec en Outaouais, Canada (2009), a Visiting Researcher at the Purdue University, USA (2001), and the Hong Kong University of Science and Technology, China (1997–1999).

His current research interests include the different phases of the life cycle of designing data management systems/applications: requirement engineering, conceptual modelling using ontologies and knowledge graphs, logical modelling, Extraction-Transformation-Loading, deployment, physical design, personalisation and recommendation. Ladjel (co)-supervised 30 Ph.D. students on different topics covering the above phases.

Óscar Pastor (Polytechnic University of Valencia, Spain)

Óscar Pastor is Full Professor and Director of the "Software Production Methods" (PROS) at the Polytechnic University of Valencia (Spain). He received his Ph.D. in 1992. Supervisor of 20 completed PhD theses and 31 completed Masters theses on topics that relate to Conceptual Modeling, he has published more than three hundred research papers in conference proceedings, journals and books, received numerous research grants from public institutions and private industry, and been keynote speaker at several conferences and workshops.

He is Chair of the ER Steering Committee (2009–10), ER Fellow since 2010, member of the SC of conferences as CAiSE, ER, ICWE, ESEM, CIbSE or RCIS, and member of over 100 Scientific Committes of top-ranked international conferences, his research activities focus on conceptual modeling, web engineering, requirements engineering, information systems, and model-based software production.

Óscar Pastor created the object-oriented, formal specification language OASIS and the corresponding software production method OO-METHOD. He led the research and development underlying CARE Technologies that was formed in 1996. CARE Technologies has created an advanced MDA-based Conceptual Model Compiler called Integranova, a tool that produces a final software product starting from a conceptual schema that represents system requirements. He is currently leading a multidisciplinary project linking Information Systems and Bioinformatics notions, oriented to designing and implementing tools for Conceptual Modeling-based interpretation of the Human Genome information.

Keynote Abstracts

Towards Green Data Management Systems

Ladjel Bellatreche ⓘ

LIAS/ISAE-ENSMA, Poitiers University,
Futuroscope, 86960 Poitiers, France
`ladjel.bellatreche@ensma.fr`

Abstract. In today's world, our life depends too much on computers. Therefore, we are forced to look at every way to save the energy of our hardware components, system software, as well as applications. Data Management Systems (DMSYSs) are at the heart of the energy new world order. The query processor is one of the DMSYS components in charge of the efficient processing of data. Studying the Energy-Efficiency of this component has become an urgent necessity. Most query optimizers minimize inputs/outputs operations and try to exploit RAM as much as possible. Unfortunately, they generally ignore energy aspects. Furthermore, many researchers have the opinion that only the OS and firmware that should manage energy, leaving DMSYSs as a second priority. In our opinion, software and hardware solutions must be integrated to maximize energy savings. This integration seems natural since query optimizers use cost models to select the best query plans and use hardware and software parameters. As scientists, we first feel obliged to motivate researchers to embrace the Energy-Efficiency of DMSYSs through a survey. Secondly, to accompany them, we propose a road-map covering the recent hardware and software solutions impacting query processors. Finally, guidelines for developing green query optimizers are widely discussed.

Conceptual Modeling and Life Engineering: Facing Data Intensive Domains Under a Common Perspective

Oscar Pastor(iD)

PROS Research Center, Universitat Politècnica de València,
Camino Vera s/n., 46022 Valencia, Spain
opastor@pros.upv.es

Abstract. Understanding the Hunan Genome is the big scientific challenge of our century. It will probably lead to a new kind of "Homo Sapiens" with new capabilities never before affordable for human being as we know them. This will be referred in the keynote as an "Homo Genius" evolution. Getting a shared understanding of the domain is a first essential task in order to manage correctly and efficiently such a complex data intensive domain. With more and more data being generated day after day with the continuous improvements of sequencing technologies, selecting the right data management strategy intended to support the design of the right software platforms that successfully attend user requirements (in terms of relevant information needs) becomes a unavoidable, crucial goal. This talk discussed how the use of a sound conceptual modeling support is fundamental in order to convert that "understanding the genome challenge" into a Life Engineering problem, where conceptual modeling, explainable artificial intelligence and data science must work together in order to provide accurate, reliable and valuable solutions, putting an special emphasis in the modern Medicine of Precision applications.

Contents

Data Integration, Conceptual Models
and Ontologies

Managing Data-Intensive Research Problem-Solving Lifecycle

Nikolay Skvortsov[✉] and Sergey Stupnikov

Institute of Informatics Problems, Federal Research Center "Computer Science and Control",
Russian Academy of Sciences, Moscow, Russia

Abstract. Problem-solving lifecycle providing provable semantic interoperability and correct reuse of data, metadata, domain knowledge, methods, and processes on different levels of consideration is proposed. It includes ontological search, data model integration, schema mapping, entity resolution, method and process reuse, hypothesis testing, and data publishing. Problems are solved according to formal domain knowledge specifications over multiple integrated resources. The semantics of every decision may be formally verified.

Keywords: Problem-solving lifecycle · Conceptual modeling · Data model unification · Ontological relevance · Schema mapping · Entity resolution · Semantics verification

1 Introduction

Preservation and accessing research data to ensure their reuse and reproducibility of research results has been a pressing issue for many years. Researchers use multiple data resources and try to make the results of their work reachable to the research community. The volume of heterogeneous data and the variety of ways to analyze it have become such that research communities no longer have the ability to spend most of their time and effort resolving data heterogeneity manually for each problem they solve.

Distributed research data repositories and tools, shared format and interface standards for supporting and managing research data are being developed. Interdisciplinary research data infrastructures are being created to provide search, access, and analysis of data of various nature. There is still no stable position on how to organize this support, which services should be offered to researchers, and how to automate and simplify the research process.

An excellent summary of previous efforts is the FAIR data principles [15], which declare forward-looking requirements for data management. Supporting data with metadata, provenance information, providing access, searching, and using semantic technologies are prerequisites for achieving data interoperability and reuse.

The FAIR data principles have become one of the main directions for actively discussed foundations for creating global interdisciplinary international research data

© Springer Nature Switzerland AG 2021
A. Sychev et al. (Eds.): DAMDID/RCDL 2020, CCIS 1427, pp. 3–18, 2021.
https://doi.org/10.1007/978-3-030-81200-3_1

infrastructures. It is clear for now that means for search and access to data, rich descriptions of data semantics, data integration, and automation of research on data are necessary. There are communities of researchers interested in achieving these goals working with data infrastructures in their disciplines. In this context, the purposes of the discussion in this paper are possible problem-solving lifecycle in data infrastructures and proposing the semantic specifications of domains as its core.

In [10] a lifecycle of research was proposed that was considered as a reflection of the FAIR data principles. The heart of this lifecycle is formal domain specifications maintained by communities working in those domains. Management of domain descriptions makes it possible to search for data relevant to the research and for methods to reuse them, then solving problems in terms of domain specifications and represent research results in the same terms. In this way, research results remain reachable for reuse in other research within the community.

This paper proposes an enhancement of the lifecycle of research problem solving which can combine both simplified accustomed approaches of research groups to data analysis, and the possibility of semantic approaches to managing heterogeneous data and formal verification of their correct usage at different stages of integration in solving problems.

The following chapters are organized as follows. Section 2 describes the state of approaches to research in groups and communities and the interoperability problems remaining in them. Section 3 tells about levels of heterogeneity and resource integration. Section 4 proposes a lifecycle of data management for problem-solving taking into account resolving data heterogeneity once for every new resource in the community and reuse of integrated resources. Sections from 5 to 8 describe in detail different stages of the lifecycle. In Section 9 conclusions about necessary services for data management during the lifecycle have been described.

2 Current Trends in Data Management for Research Problem Solving

Most disciplines have become data-intensive and use data resources to discover new knowledge from them. Different kinds of research meet certain difficulties with data management.

- Projects that use several resources to cover observations of research objects often spend the major efforts on data integration. It is necessary to apply particular resources in their stage to solve the problem. Their structures are used together with internal structures created during the problem-solving. Sometimes materialized integration technologies like ETL are used. Data resources have to be integrated, a global schema is created for them, data are transformed, identified, and combined in the representation used in the project. Virtual integration is rarely used in practice, even for dynamically changing data sources. Instead, ETL may be used to update data. Most problems require the integration of data resources or the reuse of the same structures in the new research. The developed methods if they remain accessible can be applied to other data.

- Research at the intersection of different knowledge domains occurs very frequently. It simultaneously uses data resources related to those domains. In such problems, the same difficulties arise as above, but data integration and problem-solving are more complicated because specialists in different domains have to interact, and the data resources are significantly different.
- Machine-learning-based studies use much data with known parameters of interest for training, and observational data to obtain parameters based on trained models. Statistics and clustering methods are used when data for training are not available. To solve problems with machine learning methods, it is often necessary to find high-quality data having a wide set of parameters with well-known semantics. To evaluate the possibility of using data and the quality of the built models, it is critical to know the provenance of the data, can reproduce the results. The accessible datasets in different structures have to be integrated and preprocessed to apply machine learning methods. Results are not often published. When publishing, it must be clearly stated that the results were obtained using machine learning, not observation.
- Monitoring dataflow from one or more sources used together with collected historical data for evaluating and predicting some parameters is often implemented as workflows. This type of research meets changes in data source structures and accessibility and the need to maintain the process of problem-solving.
- The modeling of research objects is used to evaluate the unobservable parameters of research objects. Hypotheses on such objects and models are generated, parameters that can be observed are calculated for models, and then the modeled data are compared to the observational data. Hypothesis testing is performed to detect laws. In such studies, it is necessary to know existing hypotheses in the domain and to experiment with combinations of them. At the same time, the experimenting process with different hypotheses can be unified and not be dependent on the domain.
- Research in critical domains requires validation of developed means. The correctness of the research may depend on the correct transformation of data during their integration, correct interpretation of data, correct application of methods to data, correct implementation of methods, etc. in which the proven correctness of the data analysis system is important.

Thus, most types of research described above require finding relevant data, have to work with heterogeneous data resources, take into account the provenance of data. They are very sensitive to the needs for semantic integration of data and to the availability of implemented methods. They spend the most effort on these needs and are interested in the reuse of correct data and method integration.

There has always been a need to tie data with methods that can process it. For instance, the object approach prevails in programming and declares the classification of data by entity types and methods as their inherent behavior. In data management and research problem solving, there may be many data sources, many ways of processing them, and many implementations of methods not limited to a single program.

In communities of research domains, libraries of methods and object classes are collected, which specialists can use for problem-solving [1, 6]. This approach suffers from the ability to include in libraries only basic methods, it is not possible to extend them with the results of various solved research problems.

There are approaches to providing search for necessary data and methods (for example web services) in registries. Usually, a search is performed by keywords or text descriptions, and the programmer needs to choose the appropriate combination of data and methods and solve heterogeneity problems. In general, using informal metadata and searching for it instead of providing semantic interoperability of data and methods is common practice.

Domain models have been discussed for a long time. And some advances in this area have been achieved in almost any research discipline community. Ontologies have been developed, unfortunately, rarely formal enough, common conceptual schemes and data formats with rich standardized metadata are used. These are usually models of subdomains most commonly used in the discipline. The usage of such models in communities becomes a requirement. This does not restrict researchers from being able to express special things in their research. However, common parts are standardized, and it ensures interoperability in communities. More specific narrow communities can formulate the specifications of their subdomains using the specifications of broader communities.

Repositories for open data are created. For interoperability, data may be published in a compatible form on the level of attributes [2]. For this purpose, a dictionary of terms and uniform requirements to corresponding values in the corresponding attributes in a uniform way are introduced. This makes it easier to use shared basic methods that work with known attributes. So the common data model helps to avoid some heterogeneity, but in fact, data remain heterogeneous both on levels of structure and values. Methods remain limited by general services.

An interesting approach is to aggregate processed data in a single container with processing methods, process specifications, documentation, links to necessary external resources, i.e., everything necessary for solving a research problem or for processing data about the research object. These containers are called research objects [5] or digital objects [16].

The problem of heterogeneity of data with significantly different nature using different data models is usually solved by choosing simple formats without a rigid structure and requirements (like JSON or CSV). The problem of significant differences in the semantics of data presented in different data models is often ignored and solved situationally.

The heterogeneity of data schemas has been sufficiently studied and is solved using special software tools for schema matching, queries to form new structures, and suggesting standard interfaces, formats, and protocols within research communities.

To automate problem-solving, process specifications controlling the sequence of data processing are predefined by specialists.

The most common approaches, such as usage of common data formats and method libraries, data search by informal descriptions, predefined automation of the research process, resource containers are not semantic-based and do not resolve data heterogeneity for research communities. Permanent efforts for data integration remain to be an obstacle for actual research.

Machine-based decision-making using analysis of data semantics, real interoperability, and automation of data management is seldom implemented, although this is the goal [8] of those who deal with the problem of FAIR principles.

3   Levels of Resource Integration

During problem-solving using data from many data sources, researchers inevitably meet data heterogeneity. These difficulties occur at different levels, including the different understanding of the domain concepts, representation of data in different data models, heterogeneity of data structure and types, and presence of different data about the same entities in different data sources. Resolving data heterogeneity is a problem that sometimes takes a lot of effort and time in research. It often becomes a cause of incorrect interpretations of data and errors in research results.

Therefore, the problem of heterogeneity of data should be paid the closest attention in data infrastructures that support research over a variety of available data resources. It is preferable to develop a lifecycle of research, which involves thorough solutions to these problems not repeatedly but based on the reuse of decisions.

Data integration involves resolving heterogeneity at each of the following levels.

- **The ontological level** of resolving heterogeneity implies semantic annotation of data with the concepts of the domain for their consistent interpretation, the possibility of searching for the necessary data resources among the available ones. In addition to the data itself, their structure, behavior specifications, data model elements, and other entities can be ontologically annotated to determine their meaning in certain domains. Annotations can be defined in terms of ontologies of various domains, either the main research area or other areas of interest. Examples of such special domains are the provenance metadata standard [4] and a data model ontology [13].
- **The data model level** defines various elements of schema definition languages, heterogeneity of which has to be considered before the differences in the structures defined in schemas themselves. Differences in the semantics of data models can be significant. It can be very embarrassing to work simultaneously with data in several data models, such as variants of object models, relational model standards, models approximated to the relational one, graph models, RDF-based models, OWL profiles, and others, without a detailed study of their semantics and the mapping rules between them.
- **The level of conceptual schemas** includes the definition of entity types in the domain, schemas that include specifications of their structures, constraints of their states, and possible behavior. Ontologies and conceptual schemas are often combined and considered at the same level of heterogeneity, ontologies are considered as a kind of data schemas. However, the problems of matching concepts of entities and homogeneous representation of data about objects have their specificities [7, 9]. Ontologies can be useful in the search for relevant data and definition of their semantics in the domain, but data ontologically relevant to the specified requirements still can be represented by different data models, use different structures and types. Different data representations defined by conceptual schemas are related to each other with the process of schema mapping. There are methods of schema matching and mapping by names, keywords, links to ontologies, machine learning, comparison of data structures and types, comparison of extents, and others. Many of them do not involve checking the correctness of schema mapping to preserve the data semantics. The correct mapping of schemas should take into account the mapping rules between data models that

are used for the integrated schemas, transforming data structure, matching constraint defined for states, and behavior of objects. As for the behavior, method and process calls should preserve the data semantics when processing them, accept correct input data, and generate results that meet the requirements of schema definitions. This is necessary for the ability to reuse not only data but method implementations as well either related to the data or shared independently.

- **The object level** includes matching domain objects in the data. The integration of schemas solves the heterogeneity in data representation, but data from different sources can contain information about the same objects. Data about an object from different sources or duplicates in one source can have different characteristics, different values, and can be valid in some contexts. Integration at the object level involves entity resolution to identify the same objects in the data, and data fusion to obtain summary information about these objects. These stages of data processing should preserve the data semantics and consistency of integration results at previous levels.

Researchers often prefer not to get involved in deep data integration and either do not state the problems that require significant integration efforts or try to use primitive data formats and interfaces. In any case, integration is performed to a limited degree, implicitly, using programming or simple software tools. This leads to resolving data heterogeneity manually in every research, but the integration result does not guarantee correctness and is not reused after finishing the research.

4   The Proposed Approach to Problem-Solving

To optimize the process of research problem-solving in data infrastructures, it is not enough to provide a search for relevant data, access them, and provide their structure metadata for integration. On the one hand, it is necessary to provide a set of services to develop a problem statement, search for data resources and methods, integrate them at all the levels described above. Verification of integration results would help to ensure its correctness and bring research to automation and machine actionability. On the other hand, having good opportunities for resource integration, it is necessary to minimize using them and maximize their reuse in many research problems being solved. For this purpose, resources are integrated for the domain, not for a single research problem.

It becomes possible when a domain community uses and supports domain models, including specific data models usually used in the domain, formal ontologies describing the domain to provide classifications of resources and semantic search for them, and conceptual schemas to represent information in the domain. Research problems are formulated using domain schemas. Resources containing data and methods should be integrated into the same domain specifications. The principles and capabilities of the conceptual approach and reuse of domain models in problem-solving were shown in [11]. In this paper, domain models are called specifications since they represent the basic requirements in terms of which research is conducted in the community. They use specification languages and they are convenient to formulate exact problem specifications based on them.

Abstract declarative domain specifications are defined. They are not intended and may not even be sufficient for implementing software systems based on them, moreover, they are independent of any implementations. But they provide domain knowledge and structures as minimum requirements that allow not only to avoid heterogeneity, and to collect well-classified collections of semantically integrated resources but also to partially automate research by delegating it to machines.

Wide communities and thematic groups are interested in the development of domain specifications for research data infrastructures. And both resource creators and users are interested in the integration of resources in such specifications. It is only necessary to organize and formalize these efforts and support them at the level of research data infrastructures. In this sense, research communities in data infrastructures should play a major role in maintaining domain specifications and registries. They are not passive user communities, but they form their domains for long-term use in research.

The idea is not to globally centralize all data representations. There are different data representations supported and traditionally used in communities, but findable mappings between them could be developed if necessary. The most general domain knowledge and data representations may indeed be unified for communities. It remains for smaller sub-communities to develop parts that relate to their specific interests. Opposing communities may appear, but their representations can be integrated if necessary.

Figure 1 shows the lifecycle of research problem-solving. The baseline approach, which is accustomed to many researchers, is depicted as a cycle, including problem specification, data resource selection, method selection, and data processing for problem-solving itself. However, each of these stages can be accompanied by a set of services to achieve interoperability at all levels and reuse data, methods, and integration results.

4.1 Specification Registries

First of all, metadata registry services are introduced, which form the basis for the interaction of all stages and activities in problem-solving. Registries are intended for the maintenance of different specifications and corresponding collections of resources of different kinds in data infrastructures. The registry of domain specifications contains:

- formal ontologies of research domains that define the concepts occurring in the research and related not only to the research objects but to methods, laws, instruments, and other notions associated with them;
- special ontologies applicable in most research domains, such as a top-level ontology, an ontology of data model elements, the standard ontology of data provenance;
- conceptual schemas that define a representation of information about domain objects agreed and accepted in the community, as well as integrity constraints, method spec-ifications, semantic annotations that link schema elements to ontological concepts to provide search.

Besides the registry of domain specifications, it is worth organizing registries for different levels of integration, including:

- data model registry for describing integrated data model elements;
- resource registry for the integration of data sources relevant to the domain research;

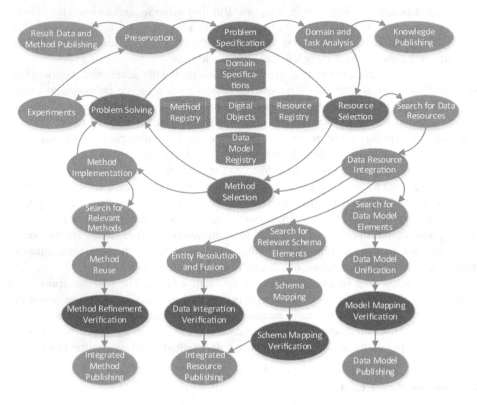

Fig. 1. The lifecycle of research problem-solving

- method registry for specifying implementations of methods and processes applicable to the domain objects.

  The registry of digital objects is used for aggregation of information related to the problem being solved, including:

- requirement specifications over the domain specifications;
- links to integrated data resources and methods used in problem-solving;
- implemented program code;
- data of intermediate results, and others.

  Any activities related to data and method selection, heterogeneity resolution and data integration, problem-solving, and providing reuse of research results focus on the analysis of specifications stored in registries and reflect the results in them.

4.2   Stages of Problem-Solving

Problem-solving following the proposed lifecycle includes the problem statement over metadata from the domain specification registry.

Data resource selection and reuse can be supported by a comprehensive set of means that provide a search for relevant data resources in the resource registry and their semantic integration at all levels, including data model integration using the data model registry, schema mapping, integration at the object level, and formal verification of integration correctness.

Selecting and reusing methods involves searching in the method registry, semantic integration of the relevant methods, and verification of the integration results. Methods can include special means for research experiments on data.

To maintain shared specifications in the research community, the final and key stage of problem-solving is publishing into metadata registries. It includes storing not only the resulting data but also the results of all stages of problem-solving. To say true, postponing the publishing at the end of the research causes researchers not to publish many important intermediate results. Thus, preferably, research plans should include publishing the results at the end of each stage such as domain analysis, resource integration, and so on. Publishing the results with due details ensures the ability of their reuse for other tasks.

In the following sections, each stage of the problem-solving lifecycle is treated separately.

5 Developing Problem Specifications

The problem-solving lifecycle normally begins with the specification of the requirements. The lifecycle allows for differing order of its stages or returning to previous stages but doesn't consider it. Problem specifications can use or be based on domain specifications, but they describe a specific goal that can be implemented. Problem statement specifications are developed as a specialization of domain specifications. Based on the requirements of the problem statement, additional requirements can be imposed on the ontological concepts.

The obtained ontological specifications of the problem, which are specializations of the domain ontology concepts, on the one hand, are used to search for relevant entity types of conceptual schemas and to specify the problem in terms of conceptual schemas. On the other hand, the same requirements in terms of ontologies are used as requests to the resource registry. Thus, the ontological level semantically links the requirements for solving the problem and the requirements for data that can potentially be used for solving it.

6 Selection and Integration of Relevant Resources

The selection and integration of data resources relevant to the problem may be divided into several stages and significantly formalized. The process begins with searching for resources in the registry. Queries could use concepts developed for the problem requirements. Source data or previous research results of the community are returned if their ontological descriptions are specializations of the queries. It should be noted that these data were included in the registry as ontologically relevant to the domain, and the search finds only resources that are relevant to the problem.

If selected resources had been published properly, they are probably integrated into the domain specifications. Such resources can be reused without an integration phase. If new resources need to be used, or resources have not been fully integrated yet, integration at the levels of data models, conceptual schemas, and objects should be applied to them.

6.1 Data Model Integration

Resources being integrated into the domain specifications can be represented in data models that were not previously used in the community. In this case, the resource data models must be mapped to a model used in the community as canonical.

Data model unification [14] creates a set of extensions of the canonical model in the data model registry that maps elements of resource data models expressing them with specific structures, data types, operations, or constraints in the canonical model. Reference schemas specifying data models express their abstract syntaxes and semantics. Model matching is similar to schema matching.

If a kind of data model ontology [13] is used in the community, the data model integration process can start with an ontological description of data model elements and search for relevant canonical model elements or extensions that already exist in the registry. It allows reusing the existing extensions for similar data model elements which include mapping rules in terms of reference schemas.

If there are no suitable extensions, rules converting data model elements should be developed and new extensions should be created. The result is saved in the data model registry and accompanied by semantic annotations in terms of the ontology.

Formal verification of data model mapping is possible. Any schema of a resource data model should refine the mapping in terms of the canonical model. Informally, the specification refinement relationship assumes that a specification can be replaced by another one without a noticeable difference to the user. To verify the refinement, the resource model specifications and the canonical model extensions can be formalized in the abstract machine notation (AMN). B-technology [3] supports semi-automated prove of the refinement relation between abstract machines.

6.2 Schema Integration

After solving the heterogeneity of data models, the conceptual schemas of data resources are integrated. Firstly, a semantic search could be performed to find and preliminarily link resource schema elements to the conceptual schemas of the domain or the problem being solved. The semantics of schema elements should be described and annotated in terms of the research domain ontologies. Some elements like classes had already been linked during the search and selection of data resources. Now the other relevant elements of schemas such as types, attributes, and methods are linked using ontological correspondences.

Schema mapping includes resolving conflicts between the representations of entities in data resources and their representations accepted in the domain or for the problem. Reducts of resource entity types that can refine the corresponding reducts of the domain schema entity types are detected. And compositions of type reducts from different resources are produced to implement whole specifications of the domain or the problem requirement entity types. The result is saved in the resource registry together with semantic annotations.

To be sure of the results of relevant resource integration, it is important to verify the correctness of the result specifications. The compositions of resource types should refine the entity types of the domain or the requirements of the problem being solved. The refinement relation between specifications is proven in the \abstract machine notation.

6.3 Resource Integration on the Level of Objects

After schema mapping and implementing requirement entity types with a set of relevant data resources, it is necessary to identify the same real-world entities among data from different resources, and correctly compile objects of the domain entity types. To this purpose, entity resolution and data fusion should be performed.

In entity resolution, object identification criteria can be defined for the domain based on the values of object attributes and relations with other objects. If the criteria are not included in the conceptual schema yet, they are implemented and can become candidates for inclusion in the domain specification of the community. Implementations of object identification criteria should not depend on specific data resources but be defined in terms of the conceptual schema of the domain.

Data fusion rules for identified entities may depend on the problem being solved. They can enrich data using attributes from different resources, collect object values by time, gather all values, aggregate them, select values with the best quality, and so on. Anyway, from a set of different data representations of the same entities single object with necessary data is produced for every entity.

The integration process at the object level can be quite complex and require verification. The semantics of identification criteria, data fusion rules, and the sequence of their application are defined in the abstract machine notation. It is necessary to prove that the application of rules preserves the requirements for the domain.

7 Selecting Methods and Solving the Problem

Problem requirement specifications include requirements for methods. Also, method descriptions are part of the domain specifications. Some implementations of methods related to specific research objects may be associated with data resources, some are separate resources such as services and libraries. In any case, after integrating data resource schemas, existing implementations of relevant methods should be selected and integrated, or new methods are developed to solve the problem.

Method selection can be started with a search in the method registry. Specifications of requirements as queries can ontologically express the semantics of actions performed by methods and the concepts of entities that methods use as input and output parameters. The search results return methods with corresponding semantic annotations of specification elements in the registry.

The reuse of selected method implementations is based on the principle of substitution taking into account subtype relations, weakening preconditions, and strengthening of postconditions. Methods in requirements without reusable implementations should be implemented, or found outside the registry and integrated.

A kind of method implementation is the development of workflows that consist of some related activities. Processes as a whole, their fragments, as well as single activities can be considered as methods to be reused. It is important to specify the semantics of workflows, subprocesses, activities, inputs, and outputs to provide their interoperability and reuse in the community.

The correctness of method integration can be formally verified. The specifications of the reused methods should refine the specifications of the requirements of the problem being solved.

7.1   Research Experiments

The process of problem-solving, in addition to applying methods and executing work-flows over the data of integrated resources, can also be described by special processes common for different domains, for instance, a process of research experimenting. Means for conducting experiments can include [12] generating hypotheses as methods that specify dependencies of research object parameters, these hypotheses should be refined with methods based on data modeling on the one hand and methods getting observational data from relevant resources on the other hand. Testing hypotheses include a comparison of modeled and observational results.

8   Data Publishing

Providing data reuse in research problem-solving requires paying special attention to the publishing result data, resources, and specifications. Thus, efforts are shifted from integrating relevant resources at the beginning of each new research to publishing results of every stage of the lifecycle. Results of integration and problem-solving are published once for further usage of integrated resources, methods, and data in a variety of research projects. For this purpose, the following kinds of meta-information are preferred to be published in domain communities.

- **Publishing domain knowledge**. In the registry of domain specifications, new ontological concepts may be defined. Any published concepts can refer to existing concepts, be embedded into hierarchies of them, and should keep the satisfiability of ontologies. A procedure for approving the acceptance of new concepts in the community can be introduced. Members of the community (both researchers and machines) undertake an ontological commitment to apply the domain concepts within interpretations defined by the theory of formal ontology. Knowledge not accepted by the community can still be used by subcommunities, groups, projects, or problems, and findable as subconcepts of domain ontology concepts.
- **Publishing extensions of the canonical data models**. The registry of data models the extensions of the canonical model and mappings of resources data models on them. Elements of published model extensions are also described in terms of the ontology of data models, if available, to allow preliminary search for relevant extensions. It is

preferable to publish formally verified extensions with the proven refinement of the specifications in the canonical model.

- **Publishing conceptual schemas of the domain**. Additions to the conceptual schemes in the domain are approved by the community. Enhancement of the structures or defining new modules is possible. Semantic annotations of schema elements are necessary to provide the search for relevant schema elements.
- **Publishing data resource schema integration results**. The registry of data resources is used to publish schemas of available resources and their mappings to the conceptual schemas of the domain. To provide the search for resources relevant to the problem being solved, their elements should be described ontologically with semantic annotations.
- **Publishing research results data**. Data resources containing the results of the research are created. They use integrated data models that have mappings to the canonical model. They have already been mapped to the conceptual schema of the domain since they are created based on the structures of the domain. Metainformation about created resources is published in the resource registry. Elements of their structures should have semantic annotations in terms of the domain ontology to make them findable. They should also be accompanied by provenance metadata to reflect the history of their creation based on source data, applied methods and instruments, authorship, data validity, versioning, and other descriptions. The results being published may contain more than just new result data being the ultimate goal of problem-solving. Source data observed by research instruments or simulated by machines, gathered from multiple sources, converted to a unified representation, selected with certain criteria, related to certain entity types, enriched, calculated based on source data, and others.
- **Publishing methods and processes**. It is important to accompany data published in problem-solving with available implementations of methods allowing reproducibility of results. In the method registry, metainformation includes formal descriptions of method semantics in the domain knowledge and semantics of input and output parameters. Provenance metadata link methods to data obtained with the application of them, refer to other methods used in implementations, establish authorship, and other information. Newly created methods, web services found outside the registry, dependencies, predicative criteria for entity resolution, hypotheses, laws, implementations in different ways, processes as methods, and methods as elements of processes may be published in the registry.
- **Publishing documents**. Documentation, reports, papers, and other documents referred to the research problems are published during problem-solving. Documents should be well described by metadata, and contain persistent references to published data, methods, and other results of problem-solving.

The procedure for publishing formal descriptions of the results of solving research problems in the domain community becomes necessary. Without it, it is impossible to search for data resources and methods, and access them for reuse in solving new problems. The content of published data can be developed and approved at the planning stage of research problem-solving projects.

9 Managing Problem-Solving

9.1 Digital Objects

Specifications generated during problem-solving are stored as digital objects that aggregate everything necessary to solve the problem and providing the accessibility and reuse of different kinds of resources. A digital object can contain:

- a data management plan for problem-solving;
- documentation;
- licenses;
- formal specifications of the requirements of the problem statement, including ontological specifications consistent with domain ontologies and conceptual specifications of the problem expressed with the use of conceptual schemas of the domain;
- source, intermediate, and result data,
- permanent global identifiers of reused resources, metadata related to their integration and provenance;
- developed method implementations or identifiers of reused methods necessary for data processing, their specifications and metadata.

Data management plans (DMP) [17] are formed in advance or during problem-solving. They define planned actions with data during the project such as requirements for data acquiring, processing to obtain new data, destroying outdated data, and publishing the resulting data.

9.2 Necessary Services for the Problem-Solving Lifecycle Implementation

In the lifecycle of research problem-solving (in Fig. 1), the main types of activities can be distinguished by their functionality and the used means.

- Activities related to providing search for relevant specifications (highlighted in green) are applied at different stages of selection and integration of resources and use formal ontologies to classify various resources and search for them using logical inference in the ontological model. The same type of activity can be attributed to publishing the results of problem-solving in registries since published resources become accessible for reuse using their specifications in terms of ontologies.
- Activities used to support data resource and method integration (highlighted in light blue) include data model unification, schema mapping, entity resolution, data fusion, and implementing or reusing methods.
- Activities for formal verification (highlighted in purple) include means for verifying the results of various integration stages using proof of the refinement relationship between specifications.

It allows concluding the necessary services supporting the lifecycle of problem-solving in data infrastructures to implement the FAIR data principles.

- **Registries** support long-term preserving and access to domain specifications and resources integrated into them, descriptions of implementations of methods applicable to domain objects, descriptions of integrated data models mapped to the canonical ones, and other metadata.
- **Metadata publishing and maintenance** services support metadata in terms of domain specifications and can work with descriptions of data models, conceptual schemas, data resources, methods, and processes.
- **Semantic search** services should preferably be based on logical inference in terms of formal ontologies and can be used for searching for integrated data models, data resources relevant to the problem, methods and processes applicable to domain objects.
- **Data integration** services support the integration of data resources at all levels, including the integration of their data models, schema mapping, entity resolution, and data fusion.
- **Formal semantics verification** services can provide formal proof of the correctness of specification mapping on the levels of data models, schemas, methods, and data objects.

10 Conclusion

The paper proposes an approach to managing the lifecycle of a research problem-solving that provides the reuse of resources and optimizes work with heterogeneous data and method resources. The main types of activities related to heterogeneous data and method integration, problem-solving, and publishing results are described. Using formal domain specifications, semantic search, formal verification of integration results, and detailed publishing the results of all stages of problem-solving allows consistent reuse of these results in domain communities for further research and avoid repeated integration of data resource and methods.

Acknowledgments. The research was carried out using the infrastructure of shared research facilities CKP "Informatics" of FRC CSC RAS [18], supported by the Russian Foundation for Basic Research (grants 19-07-01198, 18–29-22096, 18-07-01434).

References

1. The Astropy Project. https://www.astropy.org/
2. VizieR. https://vizier.u-strasbg.fr/viz-bin/VizieR
3. Abrial, J.-R.: The B-Book: Assigning Programs to Meanings. Cambridge University Press, Cambridge (1996)
4. Belhajjame, K., et al.: PROV-O: The PROV Ontology. W3C Recommendation. W3C (2013). https://www.w3.org/TR/prov-o
5. Belhajjame K., et al.: Workflow-centric research objects: a first class citizen in the scholarly discourse. In: ESWC2012 Workshop on the Future of Scholarly Communication in the Semantic Web (SePublica2012), pp. 1–12. Heraklion (2012)
6. Garrido, J., et al.: AstroTaverna: tool for scientific workflows in astronomy. Astrophysics Source Code Library (2013). ascl:1307.007

7. Kogalovsky, M.R., Kalinichenko, L.A.: Conceptual and ontological modeling in information systems. In: Programming and Computer Software, vol. 35, no. 5, pp. 241–256 (2009)
8. Mons, B., et al.: Cloudy, increasingly FAIR; revisiting the FAIR data guiding principles for the European open science cloud. Inf. Serv. Use **37**(1), 49–56 (2017). https://doi.org/10.3233/ISU-170824
9. Skvortsov, N.A.: Specificity of Ontology Mapping Approaches. Artificial Intelligence Issues. SCMAI Transactions, no. 2, pp. 183–195. Moscow (2010). (in Russian)
10. Skvortsov, N.A.:. Meaningful data interoperability and reuse among heterogeneous scientific communities. In: Kalinichenko, L., Manolopoulos, Y., Stupnikov, S., Skvortsov, N., Sukhomlin, V. (eds.) Selected Papers of the XX International Conference on Data Analytics and Management in Data Intensive Domains (DAMDID/RCDL 2018), vol. 2277, pp. 14–15, CEUR (2018). http://ceur-ws.org/Vol-2277/paper05.pdf
11. Skvortsov, N.A., et al.: Conceptual approach to astronomical problems. Astrophys. Bull. **71**(1), 114–124 (2016)
12. Skvortsov, N.A., Kalinichenko, L.A., Kovalev, D.Y.: Conceptualization of methods and experiments in data intensive research domains. Data analytics and management in data intensive domains. In: XVIII International Conference, DAMDID/RCDL 2016, Ershovo, Moscow, Russia, 11–14 October 2016, Revised Selected Papers. Communications in Computer and Information Science, vol. 706, pp. 3–17. Springer International Publishing AG (2017)
13. Skvortsov, N.A., Stupnikov, S.A.: Application of upper level ontology for mapping of information models. In: Proceedings of the Tenth Russian Conference on Digital Libraries RCDL '2008. JINR , Dubna 2008, pp. 122–127. (in Russian)
14. Stupnikov, S., Kalinichenko, L.: Extensible unifying data model design for data integration in FAIR data infrastructures. In: Manolopoulos, Y., Stupnikov, S. (eds.) DAMDID/RCDL 2018. CCIS, vol. 1003, pp. 17–36. Springer, Cham (2019). https://doi.org/10.1007/978-3-030-23584-0_2
15. Wilkinson, M., et al.: The FAIR guiding principles for scientific data management and stewardship. In: Scientific Data, vol. 3 (2016)
16. Wittenburg, P.: From persistent identifiers to digital objects to make data science more efficient. In: Data Intelligence, vol. 1, no. 1, pp. 6–21 (2019). https://doi.org/10.1162/dint_a_00004
17. Wittenburg, P., et al.: The FAIR Funder pilot programme to make it easy for funders to require and for grantees to produce FAIR Data. arXiv preprint arXiv:1902.11162 2019
18. Regulations of CKP "Informatics". http://www.frccsc.ru/ckp

Algebraic Models for Big Data and Knowledge Management

Sergey Makhortov$^{(\boxtimes)}$ (iD)

Voronezh State University, 1, Universitetskaya pl., Voronezh 394018, Russia

Abstract. When processing large volumes of information, obtaining an accurate solution often requires excessive resources or is even impossible at all. Therefore, in modern information systems, methods of artificial intelligence and inference are used, which allow to find approximate solutions with a given accuracy in an acceptable time. An important feature of such systems is the fuzzy nature of knowledge and reasoning. Knowledge bases with some simplified logic, for example, based on fuzzy production rules can serve as a basis for their construction. Effective formalism for the construction and study of data and knowledge models is provided by algebraic methods. In particular, the formal methodology for knowledge management in production-type systems is developed by the lattice-based algebraic theory of LP-structures (lattice production structures). Another achievement of the theory is the method of relevant backward inference (LP-inference), which significantly reduces the number of queries to external sources of information. This result is especially relevant when processing big data and managing large volumes of knowledge. The basis of LP-inference is the apparatus of production-logical equations. The article for the first time introduces an extended class of such equations for an algebraic model, the expressive capabilities of which cover logical inference systems with a fuzzy knowledge base. Some properties of these equations are proved that are useful for finding their solutions. Thus, the advantages of the theory of LP-structures extend to a fuzzy logical inference.

Keywords: Fuzzy production system · Relevant backward inference · Algebraic model · Fuzzy LP-structure · Production-logical equation

1 Introduction

Significantly affecting the life of modern society, information systems have become an essential component of business processes. The range of their application is constantly expanding, they are becoming more complex, aquiring global and distributed character. Due to the criticality of a number of subject areas, the information systems requirements are becoming significantly tougher, the complexity of their designing and developing is rising significantly. Accordingly, the relevance of mathematical justification of the correctness and reliability of information processing in various subject areas is growing. These areas include, in particular, the tasks of big data and knowledge management.

© Springer Nature Switzerland AG 2021
A. Sychev et al. (Eds.): DAMDID/RCDL 2020, CCIS 1427, pp. 19–26, 2021.
https://doi.org/10.1007/978-3-030-81200-3_2

When processing large volumes of information, obtaining an accurate solution often requires excessive resources or is even impossible at all. Therefore, in modern information systems, methods of artificial intelligence and inference are used, which allow to find approximate solutions with a given accuracy in an acceptable time. An important feature of such systems is represented by the fuzzy nature of knowledge and reasoning [1]. The basis for their construction can be knowledge bases with some simplified logic, for example, based on fuzzy production rules [2]. The possibilities of using intelligent systems with fuzzy rules for big data processing are described in a number of publications [3, 4].

It should be noted that production-type logic is used not only in traditional for it expert systems [5], but also in a number of other areas of computer science. This circumstance is explained by that in fact a lot of models in computer science are of production character [6], and the structures for presenting data and knowledge are often hierarchical [7]. In particular, with the help of productions it is possible to describe functional dependencies in relational databases calculated using inference based on Armstrong's axioms [8]. These dependencies play an important role in designing the structure of databases, especially «big» ones. In the article [9], it was shown for the first time that the relations of generalization and type aggregation in object-oriented programming possess the properties of production-logical inference. In [10], similar properties were noticed and used to optimize term rewriting systems.

An effective formalism for constructing and studying data and knowledge models in various subject areas is provided by algebraic methods [9–11]. In particular, the formal methodology for managing knowledge and logical inference in production type knowledge systems is developed by the lattice-based algebraic theory of LP-structures (lattice production structures) [12, 13]. It is intended for the formal solution of a number of important tasks related to production systems. These include equivalent transformations, verification, equivalent minimization of knowledge bases. Another achievement of the theory is the relevant backward inference method (LP-inference), which significantly reduces the number of calls to external information sources [14]. This result is especially relevant when processing big data and managing large volumes of knowledge. LP-inference is based on the apparatus of production-logical equations.

In this article an extended class of such equations for an algebraic model, the expressive capabilities of which cover logical inference systems with a fuzzy knowledge base, is introduced for the first time. Some of these equations properties useful for finding their solutions are proved. Thus, the advantages of the theory of LP-structures extend to a fuzzy logical inference.

It should be noted that there are many works by other authors devoted to the fuzzification of algebraic systems, including those based on lattices. A separate article can be devoted to a review of their results. However, these studies do not overlap with the subject of this work, so their bibliography is not given here.

The article consists of the following main sections. In Sect. 2, the necessary basic concepts and notations are introduced and initial mathematical results are formulated. In Sect. 3, a new apparatus of production-logical equations in a fuzzy LP-structure is introduced, their useful properties that underlie the solution method are established. In Sect. 4, the main ideas for the practical application of production-logic equations for acceleration of fuzzy backward inference (relevant FLP-inference) are discussed. In the Conclusion, the results are summed up and some of the research perspectives are indicated.

2 Foundations of the Fuzzy LP-Structures Theory

The starting for this section points of the theory of lattices, fuzzy sets, and binary relations are described, for example, in [1, 15]. Recall some of them in order to fix the notations used here.

A fuzzy set $A = (F, \mu_A)$ is determined by the membership function $\mu_A : F \rightarrow [0, 1]$ on some (ordinary) set F. The value $\mu_A(a)$ is called the degree of truth of $a \in F$. A fuzzy binary relation R on a set F is a fuzzy set of ordered pairs of elements from F with a given membership function $\mu_R : F \times F \rightarrow [0, 1][0, 1]$.

A relation R on an arbitrary set F is called reflexive, if for any $a \in F$ $\mu_R(a, a) = 1$ is true.

To model fuzzy logical inference, one can use the composition of fuzzy relations, in particular, in the classical semantics, namely – the (max-min) composition. The composition $R^2 = R \circ R$ is defined as follows:

$$\mu_{R^2}(a, c) = \max_b(\min(\mu_R(a, b), \mu_R(b, c))), \text{ where } a, b, c \in F.$$

A fuzzy binary relation R on a set F is *transitive* if for any $a, b, c \in F$ $\mu_R(a, c) \geq \min(\mu_R(a, b), \mu_R(b, c))$ is true. Thus, the property of transitivity for a relation R is equivalent to embedding $R^2 \subseteq R$ in the sense of fuzzy sets. There is a closure of an arbitrary fuzzy relation with respect to the properties of reflexivity and transitivity. A review of the algorithms of its construction is presented in [16].

The problem of finding transitive reduction is known: for a given relation R a minimal fuzzy relation R' is sought such that its transitive closure coincides with the transitive closure R. The solution of this problem is considered, in particular, in [17].

Let an atom-generated lattice $\mathbb{F}$ be given, which is the set of all finite subsets of a certain set F. To emphasize this, instead of symbols $\leq, \geq, \wedge$ and $\vee$ (accepted in the general theory of lattices [13]) we will use the symbols of the set-theoretic operations $\subseteq, \supseteq, \cap$, and $\cup$, and designate the lattice elements, as a rule, with capital letters. The exception are atoms indicated by lowercase letters.

On $\mathbb{F}$ a fuzzy binary relation R containing $\supseteq$, as well as having transitivity and distributivity, is introduced. The last property has the following semantics [18].

Definition 1. A fuzzy binary relation $R = (\mathbb{F}, \mu_R)$ is called distributive, if for any $A, B_1, B_2 \in \mathbb{F}$ $\mu_R(A, B_1 \cup B_2) \geq \min(\mu_R(A, B_1), \mu_R(A, B_2))$ is true.

A relation with the above properties (contains $\supseteq$, transitive, distributive) will be called production-logical, or, for brevity, simply logical.

Definition 2. By fuzzy LP-structure (FLP-structure) an algebraic system is meant, which is a lattice on which a fuzzy production-logical relation is given.

As noted in [18], such a structure is an algebraic model of an intellectual system of a production type with fuzzy rules (for example, similar to [5]). Lattice atoms correspond to the simplest facts, unions of atoms (lattice elements) correspond to sets of facts. The considered fuzzy logical relation on a lattice models a sets of fuzzy rules (productions). The properties of this relationship in a natural way reflect the possibilities of fuzzy inference.

For further discussion, some results related to FLP-structures, established in [18], are required.

The initial relation R specified on the lattice is usually not logical, but its logical closure $\bar{R}$ can be considered. This closure essentially models all potential inferences in a production system.

A logical closure $\bar{R}$ of a fuzzy binary relation R is the smallest logical relation containing R. Note that the inclusion relation $\supseteq$ determining the lattice is itself a logical relation, the smallest of all possible.

In [18], the existence of a logical closure for an arbitrary R was proved and its structure was described. This result allowed us to introduce the concept of equivalent relations, that is, in applications – formally equivalent fuzzy knowledge bases.

Two fuzzy relations R, P on a common lattice are called (logically) equivalent ($R \sim P$) if their logical closures coincide. An equivalent transformation of a fuzzy relation R is such a modification of its membership function ($\mu_R \rightarrow \mu_P$) that the resulting new relation P is logically equivalent to R.

A fuzzy relation on an atom-generated lattice $\mathbb{F}$ is called *canonical* if its membership function is positive only on pairs of the form (A, a), where $A \in \mathbb{F}$, a is the atom in $\mathbb{F}$. In [18], a theorem on the existence of an equivalent canonical relation for an arbitrary R was proved. In the modelled production system, the canonical relation corresponds to the set of rules of the so-called Horn type [19].

3 Logical-Production Equations in the FLP-Structure

In this section a new class of equations related to fuzzy LP-structures is introduced. The question of a method for solving these equations is considered. The process of finding a solution corresponds to the backward fuzzy logical inference in the production system. At the same time, the right-hand side of the equation reflects the hypothesis being tested, and each solution found is a set of facts. If all the facts in solution are true the hypothesis will be proven correct.

Let a fuzzy relation R on an atom-generated lattice $\mathbb{F}$ be given. Let also $\mu_R(A, B) > 0$ take place for some elements $A, B \in \mathbb{F}$. Then B can be called an image of A, and A – a preimage of B in relation R. In the FLP-structure, each element of the lattice can have many images and preimages, and with a different degree of membership (value $\mu_R(A, B)$).

For a given $B \in \mathbb{F}$, the minimal preimage with relation R is such an element $A \in \mathbb{F}$ that $\mu_R(A, B) > 0$ and A are minimal in the sense that it does not contain any other $A_1 \in \mathbb{F}$ for which $\mu_R(A_1, B) > 0$.

Definition 3. An atom $x \in \mathbb{F}$ is called initial in a fuzzy relation R if there is not any pair $A, B \in \mathbb{F}$ such that $\mu_R(A, B) > 0$, x is contained in B and is not contained in A. An element X is called initial if all its atoms are initial. A subset $\mathbb{F}_0(R)$ (sometimes denoted $\mathbb{F}_0$) consisting of all the initial elements $\mathbb{F}$ is called the initial set of lattice $\mathbb{F}$ in relation R.

The initial set $\mathbb{F}_0$ defined above forms a sublattice in $\mathbb{F}$.

Let $\bar{R}$ be a logical closure of a relation R (see Sect. 2). Taking into consideration its structure [16], it is easy to verify that the sets $\mathbb{F}_0(R)$ and $\mathbb{F}_0(\bar{R})$ coincide.

Consider the equation

$$\bar{R}(X) = B, \tag{1}$$

where $B \in \mathbb{F}$ is the given element, $X \in \mathbb{F}$ is the unknown.

Definition 4. *An approximate solution* to Eq. (1) is any preimage of an element B in $\mathbb{F}_0$ (in relation $\bar{R}$). *The solution (exact)* (1) is any minimal preimage of element B in $\mathbb{F}_0$. *A general solution* to an equation is the set of all its solutions.

Equations of the form (1) will be called *production-logical equations* in a fuzzy LP-structure.

Remark 1. By definition, the exact solution to Eq. (1) is also approximate. Besides, an approximate solution always contains at least one exact solution.

The main circumstance that creates difficulties for the solution process (1) is that usually only a relation R is given. In the modelled intelligent system, it corresponds to a given set of productions – the knowledge base. But the solution is required to be found as a preimage of the right-hand side of (1) in relation $\bar{R}$ – logical closure of R. At that, the complete construction of a logical closure is impractical, since in practice an unacceptable amount of resources will be required, both computational and in the sense of occupied memory.

An additional factor complicating not only the solution methods (1), but also the very formulation of this problem is the fuzziness of relation R. Apart from minimality of the desired preimage X required in Definition 4, it is necessary to take into account its second characteristic, namely, the value of the membership function $\mu_{\bar{R}}(X, B)$.

Apparently, the best solutions will be those which give a greater value to the membership function, respectively, in the production system – the results of the backward inference with a higher confidence coefficient. One can even try to some extent to sacrifice the accuracy of the decision in favour of increasing the confidence coefficient to practically acceptable results. A detailed study of this issue in the future may be carried with the help of multi-criteria optimization methods.

In the present work, we consider the simplest formulation of the problem for (1) as the first step of the study. We are looking for solutions on which the membership function takes positive values. However, for each solution found, this value must be calculated.

Next, we clear out the question of how the general solution of equations of the form (1) changes by their right-hand sides union. More precisely, if it is possible for a fuzzy relation R to solve several equations with simpler right-hand sides instead of the original equation.

Lemma 1. Let X_1 be a solution to an equation of the form (1) with the right-hand side B_1, and let Y_1 be a solution to an equation of the same type with the right-hand side B_2. Then $X_1 \cup Y_1$ is an approximate solution to the equation.

$$\bar{R}(X) = B_1 \cup B_2. \tag{2}$$

Proof. Since $\bar{R}$ contains the relation of inclusion, then $\mu_{\bar{R}}(X_1 \cup Y_1, X_1) = 1$ and $\mu_{\bar{R}}(X_1 \cup Y_1, Y_1) = 1$. Hence, since by the hypothesis of the lemma $\mu_{\bar{R}}(X_1, B_1) > 0$, and $\mu_{\bar{R}}(X_1, B_2) > 0$, by virtue of transitivity $\bar{R}$, we have $\mu_{\bar{R}}(X_1 \cup Y_1, B_1) > 0$ and $\mu_{\bar{R}}(X_1 \cup Y_1, B_2) > 0$. From the last two inequalities, using distributivity $\bar{R}$, we obtain $\mu_{\bar{R}}(X_1 \cup Y_1, B_1 \cup B_2) > 0$. $\qquad\square$

Theorem 1. Let $\{X_p\}$, $p \in P$ be a general solution to an equation of form (1) with the right-hand side B_1, and $\{Y_q\}$, $q \in Q$ – a general solution to an equation of the same type with the right-hand side B_2. Then the general solution to Eq. (2) is the set of all the elements of form $X_p \cup Y_q$ from which elements containing other elements of the same set are excluded.

Proof. By Lemma 1, each element $X_p \cup Y_q$ is a preimage for $B_1 \cup B_2$, that is, it contains at least one (exact) solution to (2). It remains to show that Eq. (2) has no solutions other than the form $X_p \cup Y_q$. Suppose the contrary – let some $Z \in \mathbb{F}$, being a solution to (2), not coincide with any element of the form $X_p \cup Y_q$. At that, Z cannot contain any other element of the form $X_p \cup Y_q$, otherwise it would not be a solution (it is by definition minimal). It follows that Z does not contain any X_p (or not any Y_q, which is symmetrical).

On the other hand, since $\mu_{\bar{R}}(Z, B_1 \cup B_2) > 0$ then due to $\mu_{\bar{R}}(B_1 \cup B_2, B_1) = 1$ and $\mu_{\bar{R}}(B_1 \cup B_2, B_2) = 1$, as well as transitivity $\bar{R}$, we obtain $\mu_{\bar{R}}(Z, B_1) > 0$ and $\mu_{\bar{R}}(Z, B_2) > 0$. The last inequalities mean that Z contains at least one solution to the equations with the right-hand sides B_1 and B_2. Since all these solutions are present respectively in the sets $\{X_p\}$ and, we arrive at a contradiction. $\qquad\square$

The significance of Theorem 1 is that it opens up the possibility of solving Eq. (1) by simplifying, that is, replacing it with several equations, each of which contains a lattice atom on the right side. At that, in the left part of the equations, according to Sect. 2, instead of the original relation R, an equivalent canonical relation can be considered. These transformations constitute the basic method for solving Eq. (1).

Earlier, a similar method was substantiated in detail and successfully applied to solving production-logic equations of a simpler class, namely, in the LP-structure with the usual (no fuzzy) binary relation R [13, 14]. A more detailed study of this issue in our case is beyond the scope of one article and therefore is planned for publication in a separate work.

4   On Fuzzy LP-Inference and Relevance Indicators

Let us present some ideas for the practical application of the apparatus of production-logical equations for optimizing fuzzy backward inference (relevant FLP-inference).

As is known [14], the strategy of relevant inference is aimed at minimizing the number of slowly executed queries (to a database or an interactive user). Whenever possible, requests should be consistent with those facts that are really necessary in the inference. A negative answer to a single query excludes subsequent queries about the elements of a related subset of facts (in the model – preimage of the right-hand side of Eq. (1)). Besides, with LP-inference, preference is given to testing sets of facts (preimages) of minimum cardinality.

A backward inference based on the solution of the equations begins with the construction of all minimal initial preimages in the LP-structure for the atoms corresponding to the object of expertize – the hypothesis. Further, in the constructed set, it is enough to find that preimage that contains only true facts, after which we can immediately conclude that the hypothesis is true.

The easiest way in this direction is to look at the images in succession, asking the user questions about the facts corresponding to the initial atoms (or turn to the database for them). This approach gives some advantages – only minimal preimages are investigated. You can also previously exclude "contradictory" preimages from the process, that is those reflecting incompatible facts from the point of view of the subject area.

However, there is a more efficient way – prior viewing of preimages containing the values of the most "relevant" atoms. Such are primarily the atoms that are present in the maximum number of built preimages. Then the only negative answer to the question posed excludes from consideration immediately a large number of preimages, which accordingly accelerates the study. The second indicator of the relevance of the tested atom is its presence in the preimages of minimum cardinality. Thus, preference is given to those preimages, the verification of the truth of which will require fewer questions to the user (or database queries).

Combining these two relevance indicators, you can achieve results that significantly exceed the capabilities of a standard inference machine. As experiments show [14], using the relevant LP-inference, it is possible to achieve a decrease in the number of executed slow queries by an average of 15–20%.

The fuzzy nature of the knowledge base creates additional difficulties in organizing a relevant logical inference. One of the goals of modeling such a system is to effectively calculate the truth degree of the hypothesis being tested (in the model – the membership function value of the initial preimage). Thus, when developing a strategy for fuzzy relevant LP-inference (FLP-inference), along with the mentioned above two parameters, it is necessary to take into account additional relevance indicators characterizing the fuzziness of the rules and the degree of truth of the facts. This circumstance complicates the task of optimizing the inference, but makes it much more interesting.

5   Conclusion

In the present work, the apparatus of production-logical equations in a generalized LP-structure is defined and investigated, which expands the scope of application of this algebraic theory to fuzzy intelligent systems of production type.

The properties of these equations are established, which open up possibilities for the detailed development of methods for their solution. Finding a solution to the production-logical equation corresponds to the backward logical inference in an intelligent system with fuzzy rules. The presented theorem has a simple proof, but it creates a theoretical basis for further progress in the field of optimization of fuzzy inference.

In particular, the next research steps in this direction are the detailed development of a method for solving equations and subsequently – the development of strategies for using additional relevance indicators of the FLP-inference associated with fuzzy nature of knowledge bases.

The theory being developed has prospects for practical application for building the architecture of intelligent systems, knowledge and logical inference management in various subject areas, for example, in problems of big data classification [4]. Therefore, one of the next steps is software implementation of prototype of the corresponding instrumental system.

Acknowledgments. The reported study was supported by RFBR project 19-07-00037.

References

1. Piegat, A.: Fuzzy Modeling and Control. Springer-Verlag, Heidelberg (2001)
2. Ozdemir, Y., Alcan, P., Basligil, H., Dokuz, C.: Just-in time production system using fuzzy logic approach and a simulation application. Adv. Mater. Res. **445**, 1029–1034 (2012). https://doi.org/10.4028/www.scientific.net/AMR.445.1029
3. Fernandez, A., Carmona, C.J., Del Jesus, M.J., Herrera, F.: A view on fuzzy systems for big data: progress and opportunities. Int. J. Comput. Intell. Syst. **9**(1), 69–80 (2016)
4. Elkano, M., Galar, M., Sanz, J., Bustince, H.: CHIBD: A fuzzy rule-based classification system for big data classification problems. Fuzzy Sets Syst. **348**, 75–101 (2018)
5. Buchanan, B., Shortliffe, E: Rule-Based Expert Systems: The MYCIN Experiments of the Stanford Heuristic Programming Project. Addison-Wesley (1984)
6. Maciol, A.: An application of rule-based tool in attributive logic for business rules modeling. Exp. Syst. Appl. **34**(3), 1825–1836 (2008)
7. Oles, F.: An application of lattice theory to knowledge representation. Theor. Comput. Sci. **249**(1), 163–196 (2000)
8. Armstrong, W.: Dependency structures of database relationships. In: Proceedings of IFIP Congress 1974, pp. 580–583. North-Holland, Amsterdam (1974)
9. Makhortov, S.: LP structures on type lattices and some refactoring problems. Prog. Comput. Softw. **35**, 183–189 (2009). https://doi.org/10.1134/S0361768809040021
10. Makhortov, S.: An algebraic approach to the study and optimization of the set of rules of a conditional rewrite system. J. Phys.: Conf. Ser. **973**(1), 12066.1–12066.8 (2018). https://doi.org/10.1088/1742-6596/973/1/012066
11. Hájek, P., Valdes, J.: A generalized algebraic approach to uncertainty processing in rule-based expert systems (dempsteroids). Comput. Artif. Intell. **10**(1), 29–42 (1991)
12. Makhortov, S.: Production-logic relations on complete lattices. Autom. Remote. Control. **73**, 1937–1943 (2012). https://doi.org/10.1134/S0005117912110161
13. Makhortov, S., Bolotova, S.: An algebraic model of the production type distributed intelligent system. J. Phys.: Conf. Ser. **1203**(1), 12045.1–12045.8 (2019). https://doi.org/10.1088/1742-6596/1203/1/012045
14. Bolotova, S., Trofimenko, E., Leschinskaya, M.: Implementing and analyzing the multi-threaded LP-inference. J. Phys.: Conf. Ser. **973**(1), 12065.1–12065.12 (2018). https://doi.org/10.1088/1742-6596/973/1/012065
15. Birkhoff, G.: Lattice theory. Am. Math. Soc. Providence, 418 p. (1967)
16. Garmendia, L., Del Campo, R., López, V., Recasens, J.: An algorithm to compute the transitive closure, a transitive approximation and a transitive opening of a fuzzy proximity. Mathware Soft Comput. **16**, 175–191 (2009)
17. Hashimoto, H.: Reduction of a nilpotent fuzzy matrix. Inf. Sci. **27**, 233–243 (1982)
18. Makhortov, S.: An algebraic model of the intelligent system with fuzzy rules. Programmnaya Ingeneria **10**(11–12), 457–463 (2019). https://doi.org/10.17587/prin.10.457-463. (in Russian)
19. Hammer, P., Kogan, A.: Optimal compression of propositional horn knowledge bases: complexity and approximation. Artif. Intell. **64**(1), 131–145 (1993)

A Cloud-Native Serverless Approach for Implementation of Batch Extract-Load Processes in Data Lakes

Anton Bryzgalov[1] and Sergey Stupnikov[2]($\boxtimes$)

[1] Lomonosov Moscow State University, Moscow, Russia
[2] Institute of Informatics Problems, Federal Research Center "Computer Science and Control" of the Russian Academy of Sciences, Moscow, Russia
sstupnikov@ipiran.ru

Abstract. The paper presents an approach to deal with batch extract-load processes for cloud data lakes. The approach combines multiple data ingestion techniques, provides advanced failover strategies and adopts cloud-native implementation. The suggested approach. The prototype implementation utilizes Amazon Web Services platform and is based on its serverless features. The approach can be implemented also using other cloud platforms like Google Cloud Platform or Microsoft Azure.

Keywords: Data ingestion · Serverless computing · Cloud-native applications

1 Introduction

Data integration stands for the process of creating a unified view on the basis of multiple data sources. Despite the fact that there is no universal approach to implement a data integration process, the existing solutions have common elements, such as data sources network, a master server, and clients sending queries to the master server for desired data. The query result is a consolidated, cohesive data set which is compiled by the master server from internal and external sources and is returned back to the client.

One can distinguish two ways for retrieving the required data, these are *virtual* and *materialized* integration approaches [1].

The virtual integration presumes the data are stored in the local sources and retrieved on demand. In such a way a user always get an up-to-date answer to the query. A query over the virtual global schema is rewritten into the set of subqueries that are transferred to the local sources, transformed into their query languages and executed. The data retrieved from sources have to be combined appropriately to form final answer to the initial query.

Materialized data integration paradigm requires the creation of a physical integrated data repository. One of the most widely used materialized integration techniques is *data warehousing*. A data warehouse can be considered as a set of materialized views over the operational information sources of an organization, designed to provide support for data analysis and management's decisions [2].

A. Sychev et al. (Eds.): DAMDID/RCDL 2020, CCIS 1427, pp. 27–42, 2021.
https://doi.org/10.1007/978-3-030-81200-3_3

To manage data on their way from sources to the end user, warehouses have to be accompanied with so-called Extract, Transform, and Load (ETL) systems. In a dynamic environment, one must perform ETL-processes periodically (say once a day or once a week), thereby building up a history of the enterprise [1].

Since a data warehouse is a central repository for reporting and analytical purposes, the extraction of data from various heterogeneous sources and loading them into a target data warehouse plays a very important role for enterprise data management. There are different approaches that can be used for extracting data from various heterogeneous source systems and loading these data into an Enterprise Data Warehouse. Mostly used are ETL (Extract-Transform-Load), and ELT (Extract-Load-Transform). Those abbreviations represent the order in which the data from source are processed.

Nowadays a data warehouse is often built over a *data lake*. A data lake approach involves massively scalable infrastructure which stores huge amounts of the data minimally transformed comparing to the data sources. A data lake is usually populated using EL (Extract and Load) approach and the transformation is executed right after the data is migrated from the data lake to a data warehouse.

Both ETL and ELT processes can be represented in a form of a *workflow*. That is why *Workflow Management System* (WfMS) is often considered as a part of the enterprise data infrastructure. Another important concern for modern enterprise solutions is the usage of public clouds which can dramatically reduce the infrastructure and management costs. However, to use the cloud resources efficiently an application has to meet the cloud-native design principles. Maximum utilization of cloud resources can be achieved by applying the serverless paradigm which means that the resources are allocated only at the time of actual code execution and teared down when the execution is finished.

The serverless paradigm is applicable for the Workflow Management Systems (WfMSs). Such serverless WfMSs have two types of nodes: the management nodes and the workers nodes. The workers can be run in a serverless manner being set up at the time when a task to be executed appears. A management node schedules the tasks for execution and monitors dependencies between them. The use case of *enterprise batch EL* (data are loaded from sources into a data lake in large portions, but loading process is not continuous and presumes gaps between batch loads) allows management node to be switched on at the beginning of EL routine to populate the tasks queue and then to be shut down while the workers execute the tasks on their own.

Serverless approach allows us to be billed only for the actual utilization of resources during their runtime, so the cloud infrastructure costs are reduced, especially for the use case of batch EL.

The present paper advocates new approach to deal with batch EL processes for data lakes. From a high-level perspective, the approach includes three main steps: data extraction tasks execution, data partitioning and deduplication. As a result, the data is stored in a persistent storage and can be further used by outer data processing infrastructure. The proposed approach is aimed to fit use cases which are characterized by (1) discontinuous batch jobs (2) that are executed in a cloud infrastructure. EL processes are run according to a schedule and execution time of a single run is less than an interval between the sequential executions. This causes existence of idle periods for infrastructure which leads to underutilization of resources and leaves an opportunity to apply serverless app-roach. Utilization of resources is increased by shutting them down after an execution has finished. The approach is implemented on the basis of Amazon Web Services (AWS) cloud infrastructure.

The rest of the paper is organized as follows. Section 2 describes the state of the art in the area of batch data ingestion in cloud data lakes. Section 3 provides a review of related works including classification of approaches and comparing their advantages, disadvan-tages and common features. Section 4 describes software architecture for the proposed approach and the implementation issues of individual components of the architecture. Section 5 illustrates evaluation of the implementation. Final discussion and conclusions form Sect. 6.

2 Batch Data Ingestion in Cloud Data Lakes

Two common approaches to populate data warehouses with data from heterogeneous sources can be distinguished: ETL and ELT.

ETL (Extract-Transform-Load) approach to data warehouse development is the tra-ditional and widely accepted one. Data are "extracted" from the data sources using a data extraction tool via whatever data connectivity is available. After that data are trans-formed using a series of transformation routines. This transformation process is largely dictated by format of the output data. Data quality and integrity checking are performed as a part of the transformation process, and data correction actions are built into the process. Transformations and integrity checking are performed within so called *data staging database*. Finally, once the data is converted into the target format, they are then loaded into the data warehouse [3].

Data storage costs are becoming cheaper with the lapse of time. This allows us to perform analysis over larger amount of data with less investments. But ETL is not a uni-versal approach for any business and especially for big data management. ETL requires the data to be loaded only after the transformation is applied. So the transformation may lead to loss of potentially valuable data. One of the approaches that addresses this challenge is Extract, Load and Transform (ELT).

The basic idea of ELT approach is to load data immediately after it is extracted. The transformation should only be applied after the data is persistently stored. This allows to keep a validated and cleaned offline copy of the source data. Both the extraction and loading processes can be isolated from the transformation [3]. ELT has following advantages compared to ETL: (1) data sources can be added in a flexible way; (2) aggregation can be applied multiple times on same raw data; (3) transformation process can be adapted for different data; (4) process of implementation is sped-up [4].

Modern enterprise analytical solutions often implement hybrid data infrastructure architectures combining data warehouse with data lakes. Using a data lake allows to keep a vast amount of raw data in its native format («as is»). The data lakes are implemented using massively scalable infrastructure and able to handle large and quickly arriving volumes of unstructured and semistructured data. Data Warehouses are usually in power of processing structured data only. Processing systems can also ingest data from the data lake without compromising the data structure [5].

Hybrid solutions are preferred to standalone warehouse solutions due to former flexibility in terms of heterogeneity of stored data [6] and hardware costs [7]. The data lake-based data platform can be evolved further to support analytical needs of the business. Many analytical usage scenarios can be considered for the data lakes [8] but they are beyond the scope of this work. In this paper we focus on data fail-safe ingestion infrastructure design.

ELT applications are typically modeled as workflows including tasks, data elements, control sequences and data dependencies. Workflow management systems (WfMS) are responsible for managing and executing these workflows. Typical architecture of WfMS consists of three major parts: (a) the user interface, (b) the core, and (c) plug-ins [9]. Plug-in components are responsible for interacting with various resources while the core components are responsible for workflow orchestration and scheduling.

Cloud computing services are a good choice to process and manage large amounts of data which require the use of a distributed computational environment which become possible due to recent progress in virtualization technologies. Existing cloud resource pools can be used (and often cheaper) for on-demand and scalable scientific computing. Cloud services also make scalability of the service more maintainable by real-time provisioning of resources to meet the dynamics of changing application requirements implemented based on the Infrastructure-as-a-Service (IaaS) and Platform-as-a-Service (PaaS) solutions. On the opposite, when the demand is low then the underutilized resources are shut down [10]. However, effective usage of cloud resources requires the application to match a set of principles to be identified as a cloud-native application. This means it has to be adapted according to design patterns that have been used in many successful cloud applications [11].

One of the most recent cloud application development approaches is *serverless* one [12]. Serverless computing describes a cloud-native programming model and architecture where small code snippets are executed without any control over the resources on which the code runs. The servers are still a part of an infrastructure, but they are not maintainted by the developer. Most of the operational concerns such as resource provisioning, monitoring, maintenance, scalability, and fault-tolerance are left to the cloud provider [13].

Serverless paradigm may overlap with other cloud-related terms such as PaaS and Software-as-a-Service (SaaS). Following the definitions of the varying levels of developer control over the cloud infrastructure we may define serverless approach as a Function-as-a-Service model. Since the IaaS model allows the developer to control both the application code and operating infrastructure, the PaaS and SaaS models allow the developer Software-as-a-Service only to control the application itself without any control over the infrastructure, the Function-as-a-Service model gives the developer control over the code they deploy into the Cloud with a limitation that code has to be written in the form of stateless functions. All of the operational aspects of deployment, maintenance, scalability and fault-tolerance of that code are left to a cloud provider. In particular, the code may be scaled to zero where no servers are actually running when the user's function code is not used, and there is no cost to the user.

While the serverless paradigm allows business to spend literally no developer time on any infrastructure operations and has a significant economic impact, the approach requires some architectural efforts devoted to ELT processing.

Big data analytics systems review [14] shows that current ELT and ETL solutions are often based on massive parallel processing architectures and are tightly connected with the hardware they are based on. Purely serverless EL solutions are not known.

Another important concern is that enterprise data routines are usually implemented in form of batch ELT or ETL processes which are executed during a condensed time period [15]. This allows us to narrow the target systems class to serving the purposes of discontinuous batch data ingestion processes.

In this paper we present an approach to develop a serverless batch EL solution. Serverless implementations of EL-systems have the following advantages [16]: (1) easier operations including monitoring the health of worker nodes, (2) higher resources utilization, meaning pure pay-as-you-go cost model, and (3) highly scalable solution, as far as serverless platforms allows gaining much advantages of autoscaling.

3 Related Works

This section contains an overview of several EL-related solutions and approaches as well as WfMSs used to represent ETL-ELT pipelines allowing us to highlight principal features of EL steps.

ELTA approach in [4] stands for Extract, Load, Transform and Analyze. The authors define the ELTA term as follows: «a process called Extract enables data extraction from the heterogeneous sources in heterogeneous formats (transactional data, machine generated data etc.); process Load provides ability to store data inside storage system; process named Transform provides ability to transform data from raw state (a) on demand and (b) according to the needs of decision making process; Analyze phase makes business users efficiently utilize preprocessed data to understand enterprise behavior through analysis». Extract and Load Processes in ELTA approach are based on the metadata defined in the Zachman Enterprise Architecture framework [17], the IT users can extract all necessary for business information from heterogeneous data sources and load it in a big data storage.

Another E-LT approach is proposed in [18] and is called E-Hub. According to the authors, E-Hubs are opposite to the service providers, which are split into Business Service Providers (BSPs) and Application Service Providers (ASPs). Those entities differ in the business-related application they are designed for: ASPs provide convenient access to IT infrastructure and operational expertise without addressing business or inter-company market exchange requirements, whereas BSPs address business process issues without enough access over infrastructure. E-Hubs are neutral Internet-based intermediaries that focus on specific industry verticals or specific business processes which provide secure trading environments to link with external buyers and suppliers. The implementation of proposed E-Hub architecture is based on Oracle Data Integrator. The architecture is built on top of the repository containing configuration of the entire IT infrastructure, the metadata for all applications, projects, scenarios, execution logs, defined by the administrators. Relational OLTP database can be used as the repository.

Authors of [19] present an ETL approach for near real-time data ingestion with focus on every step including Extraction. The process of extraction involves profiling data dimensional attributes from separate databases, capturing data changes and getting data from the source location. Streaming data such as network traffics, click streams and sensors are fleeting, constantly changing and continuous. The approach challenges the way data changes are captured for frequent update and loaded without becoming overloaded and disrupting normal operational activities.

Workflow management systems take care of defining, managing and executing workflows which can be used to decently represent ETL-ELT pipelines. A workflow models a process as a series of steps (tasks) that simplifies the complexity of execution and management of applications [9]. A workflow model (or specification) includes task definition and workflow structure definition. Graph-based modeling tools allow the definition of an arbitrary workflow using a few basic graph elements.

Popular workflow systems for scientific applications are DAGMan (Directed Acyclic Graph MANager based on Pegasus WfMS) [20, 21], Pegasus [22], Kepler [23], and Taverna workbench [24]. Kepler is a WfMS that supporting Web-service-based workflows and implementing an actor-oriented design approach for composing workflows. The Taverna is a workbench which automates experimental methods. It allows to integrate various services, including WSDL-based single operation Web services, into workflows.

Large-scale applications are implemented in the form of workflows with many precedence-constrained jobs, e.g., Montage [25] and LIGO [26]. As the number and size of the input and output data grow and precedence constraints become to be more interconnected, the complexity of such applications becomes unmaintainable. Pegasus [27] and Kepler are a popular solutions for scientific-oriented workloads to manage the execution of the workflows on clusters and grids. Using such WfMS requires to meet operational concerns including setup and configuration of the clusters consisting of physical servers or virtual machines. Setting up and configuring a large-scale cluster for a large-scale application often becomes a challenging task, especially for researchers outside the field of high-performance computing (HPC).

Serious resource underutilization is a common problem for large-scale clusters, primarily due to the complex precedence constraints among the various jobs in the workflow. That is why serverless WfMSs are being developed.

DEWE system addresses problems of (1) execution of large-scale scientific workflows and (2) resource underutilization. The hybrid (FaaService and dedicated/local clusters) job dispatching approach takes resource consumption patterns of different phases of workflow execution into account [28]. Management node application is run on a server. The jobs dependencies are built according to the workflow definition and stored into a data structure. As soon as all the job dependencies are matched, the job is eligible to run and is published to a job queue. The queue is polled by a job handler for execution. DEWE introduces two job handlers: FaaS and cluster-based. Since the FaaS execution environment has an execution duration limit, only short-running jobs can be run by the FaaS handler. A set of long-running jobs is defined for the workflow. Estimation of the execution time of a particular job can be based on module testing, previous experiences or the time and space complexity of the algorithm. If a job is expected to finish execution within the FaaS timeout, it is published into a common job queue where it can be picked by either FaaS handler or a cluster-based handler, otherwise it is published into a specific job queue for long-running jobs which is polled by cluster-based handlers only [28].

4 Serverless Batch Extract-Load System Architecture and Lifecycle

This section describes a configurable architecture which deals with batch EL processes in a cloud infrastructure. The system extracts the data from the external data sources and stores it to a data lake.

The proposed overall architecture of the EL WfMS is illustrated on the Fig. 1. The system operates with tasks which are atomic execution units and describe single portions of data to be managed.

The *task lifecycle* begins from its creation by *tasks creator* component. Tasks creator reads the configuration file provided by the user, fetches credentials from *credentials storage* and sends the requests to the source system to populate the *tasks queue* with the tasks description. The dashed arrows represent the flow of the tasks properties requests. External data *sources* contain the data as well as the metadata converted to the tasks properties. Dotted green arrows represent the responses to the credentials requests from executing components: they are moved from the credentials storage to the executing components which directly access the data sources.

Task «flows» through all the executing components as it is shown via white thick arrows. After the tasks queue is populated the tasks creator component is shut down and the *tasks executor* component (yellow badge) starts to execute tasks by sending requests to the source systems and storing the received data to the storage. Solid thin arrows represent the flow of the tasks data. Thin arrows colors differ the executing component they correspond to. Dark orange badges represent the different data locations within the data storage devoted to different storage purposes. The tasks executor stores the downloaded data to the *data landing zone*.

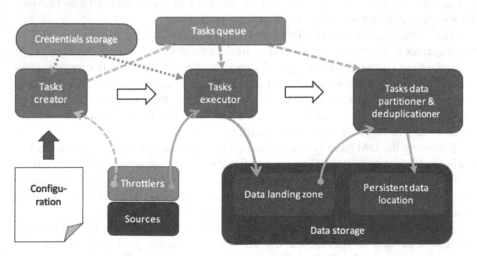

Fig. 1. Extract-load system architectural scheme

The tasks creator and tasks executor have to send requests to the source system with respect of the source system rate limits. These can be achieved by throttling the requests using the local *throttlers* which are components aimed to avoid violating the sources rate limits. If the rate limits are violated the source system may ban the requester for a longer period which will lead to increased job execution time and will increase the cloud infrastructure costs while the system will be idle.

Tasks executor stores the data to the landing zone. When all the tasks are executed then the tasks executor component stops its work and the tasks data partitioner and deduplicationer starts to merge the data from the landing zone to the persistent data location. The partitioning and deduplication are based on tasks metadata which is saved in the tasks queue. The tasks are supposed to have some identifiers which identify them in the tasks queue to retrieve the task description. The deduplication and partitioning techniques are out of the scope of current work.

The tasks creator and tasks executor components are serverless and therefore have to store all the execution status-related data in an external storage to provide fault tolerance.

As soon as the proposed EL process is supposed to be a part of an independent WfMS the full EL routine can be rerun impartibly by the managing WfMS in case of any failure. However, these may cause a great redundancy in the total number of requests to the source systems. That is why more fine-grained internal fault tolerance mechanism if preferable.

According to [29] a WfMS can be divided into three functional areas: Buildtime, Runtime Control, and Runtime Interactions. The Buildtime functions support the definition and modeling of workflow processes. The Runtime Control functions handle a process's execution, and the Runtime Interactions provide interfaces with users and applications. According to functions descriptions tasks creator component of the proposed architecture corresponds to the Buildtime and tasks executor—to the Runtime Control.

Runtime Control has two aspects: persistent storage and process navigation. Persistent storage lets the system recover from failures without losing data and also provides the means to maintain an audit trail of process execution.

This is the place where DAGMan Design Principles [30] can be applied. One of the principles is absence of persistent state of the executor component: its runtime state has to be built entirely from its input. As soon as the serverless runtime environment is considered to be ephemeral and destructible, the tasks queue becomes the persistent storage for the Runtime Control. To provide fault tolerance the tasks queue has to contain comprehensive data on the task execution status whereas the tasks executor process becomes responsible for proper status handling and updates.

Following the DAGMan principles and architectural design we come with the following task lifecycle scheme. The task is initiated with status «pending». When the tasks queue is populated by the tasks creator component all of the tasks are marked as «pending». Then the tasks creator component is turned off and all the underlying resources are destructed. The tasks definitions are stored to the tasks queue.

As the next step the tasks executor component starts to run. It may run multiple workers in parallel to increase the overall system execution pool. The worker code implements the task lifecycle logic which is expressed in a following pseudocode:

Algorithm. Running task lifecycle in a worker — *runTasks(tasksStorage, workersService, currentWorker, concurrencyLimit, tasksDataStorage, taskResetLimit)*
Input:
tasksStorage — a wrapper for tasks definitions and statuses storage,
workersService — a wrapper of a workers monitoring service,
currentWorker — the worker which runs the code,
concurrencyLimit — a maximum number of concurrent tasks executed by the worker,
tasksDataStorage — a wrapper for the data landing zone,
taskResetLimit — a limit of times the task can be reset from «running» to «pending»

```
WHILE tasksStorage.hasTasks(status IS "pending" OR status IS "running") DO
    runningTasks := tasksStorage.getTasks(status IS "running")
    activeWorkers := workersService.getWorkers(active IS TRUE)
    FOR EACH runningTask IN runningTasks DO
        IF runningTask.worker NOT IN activeWorkers:
            IF runningTask.resetCounter < taskResetLimit THEN
                runningTask.status := "pending"
                runningTask.resetCounter := runningTask.resetCounter + 1
            ELSE
                runningTask.status := "exhausted"
            ENDIF
        ENDIF
    ENDFOR


    allPendingTasks := tasksStorage.getTasks(status IS "pending")
    FOR EACH pendingTask IN pendingTasks DO
        tasksStorage.beginTransaction()
        IF pendingTask.status IS "pending" THEN
            pendingTask.status := "running"
            pendingTask.worker := currentWorker
        ENDIF
        tasksStorage.endTransaction()
        currentWorkerTasks := tasksStorage.getTasks(worker IS currentWorker)
        IF currentWorkerTasks.count() == concurrencyLimit THEN
            BREAK
        ENDIF
    ENDFOR


    currentWorkerTasks := tasksStorage.getTasks(worker IS currentWorker)
    FOR EACH task IN currentWorkerTasks DO
        TRY
            taskData := downloadTaskData(task)
            tasksDataStorage.storeTaskData(task, taskData)
            task.status := "completed"
        EXCEPT error
            task.status := "failed"
    ENDFOR
ENDWHILE
```

Every worker fetches the tasks from the tasks queue by marking them as «running» and setting the taken tasks runner identifier property to its own (worker) identifier. The running worker is treated as an active one. The worker can start running only the tasks which at the time when queue is polled have either «pending» status or «running» but with inactive worker (means the worker's runtime has been destructed due to execution error).

To avoid infinite execution of the tasks which constantly destruct the workers runtime environment every task has a counter which represents the number of times when it was reset from the running state of an inactive worker. The counter has to have a limit (*taskResetLimit*): if the task is reset more times than it is allowed the task is marked as «exhausted». Those tasks require manual incident handling.

In a successful scenario as soon as the worker has downloaded the task's data from the source system and uploaded it to the data landing zone it marks the task as «completed». If the request to the data source system was unsuccessful then the task is marked as «failed». As soon as all the tasks are either completed, failed or exhausted the tasks executor stops.

The above algorithm implements a per-task finite state machine presented on the Fig. 2. Every task can be moved from *pending* state to *running* state and back only a limited (not more that *taskResetLimit*) number of times and after that it is moved to one of three final states (*completed, failed* or *exhausted*). So every initially pending task is processed through a finite number of steps and the overall time complexity of the loop-based algorithm is $O(N)$, where N is the number of initially pending tasks.

The tasks execution stage is followed by data partitioning and deduplication. The goal is to migrate the data from the landing zone to the persistent storage area. This requires the newly fetched data to be partitioned according to the target partitioning schema. Proper data partitioning is crucial for modern query engines as they much rely on data skipping as a promising technique to reduce the need to access the data [31]. Data skipping can be achieved maintaining some metadata for each block of tuples.

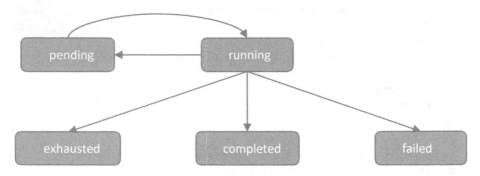

Fig. 2. Task lifecycle finite state machine

This way a query may skip a data block if the metadata indicates that the block does not contain relevant data.

The effectiveness of data skipping, however, depends on how well the blocking scheme matches the query filters. As far as the tasks definitions storage handles the tasks definitions it may be efficiently used to determine which partition the newly fetched data belongs to. The data partitioning techniques depend on the target use case. As an example of the most common technique the prototype implements the partitioning based on the date dimension [32]. This approach can be successfully applied to the proposed architecture if the data source supports date as a parameter of the request. And this is, in turn, a common case for the enterprise data sources.

Another important concern at the data migration to the persistent storage is deduplication. There are multiple deduplication techniques are available and the choice has to be driven by the application domain. The old data can be either versioned or overwritten by the newly fetched one.

Deduplication is a well-known method to optimize storage capacity [33]. Deduplication is common in the field of Data Warehousing. The data deduplication can be performed either based on the data itself using hashing or delta-encoding techniques or based on the metadata. As far as the tasks queue contains the tasks definitions they can be treated as the metadata and used for data deduplication in a following way: if the tasks metadata describes the same portion of data the newly fetched data may replace the old version. The similarity of the data presented can be tested using one of the entity resolution techniques [34]. The prototype uses an approach based on the metadata with exact tasks definitions attributes matching.

However, when the persistent data storage is located in the Data Lake, multiple versions of the same data can be stored to provide the data change timeline report as the storage is considered to be cheap enough. This way the tasks which are considered to represent the same portion of the data, but different versions, can be stored together and labelled differently.

The described architecture is implemented in a form of a serverless data application prototype based on Amazon Web Services (AWS). The correspondences between the cloud services and the architecture components are presented in Table 1.

Table 1. Cloud services representing architectural components

Component	Cloud service
Tasks creator	AWS Lambda
Tasks executor	AWS Batch
Tasks partitioner & deduplicationer	AWS Batch
Tasks queue	AWS Dynamo DB
Throttlers	AWS Dynamo DB
Credentials storage	AWS Secrets Manager
Data storage	AWS S3

The execution is triggered by AWS CloudWatch. The service triggers execution of the state machine inside AWS Step Functions which handles the order of the tasks creator, tasks executor and tasks partitioner & deduplicationer components execution.

5 Evaluation

The approach is evaluated against the implementation of the prototype based on Amazon Web Services (AWS) as a cloud provider.

The prototype has been evaluated against an enterprise EL process involving Google Analytics as a source system. Google Analytics tracks user activity on web pages. The activity is identified by user's surrogate identifier and identifier of a tracker. The data in Google Analytics is requested per view. The view represents an aggregated slice of a detailed data collected by a single tracker.

Google Analytics responses contain tabular-like data in JSON format. Tabular-like means that the response JSON object contains properties describing the data in tabular form: *columnHeaders* and *rows*. The latter contains the rows with the actual data. All other properties of the response object contain the metadata of the result. The response data sample is presented in the Fig. 3.

Google Analytics has three rate limiting rules: (1) no more than 10 requests per second are allowed from a single IP address, (2) no more than a configurable number (up to 1,000) of requests per 100 s per tracker, and (3) no more than 10 requests per second per view. The strictest limit is a per-IP one which globally limits number of requests to 10 per second. Single response contains up to 10,000 rows which is a configurable request parameter.

```
{
  "query": {
    "start-date": "2019-01-01", "end-date": "2019-01-02",
    "dimensions": ["sessionId", "hour"], "metrics": ["views", "clicks"],
    "start-index": 10001, "max-results": 10000, … // other properties
  }
  "columnHeaders": [
    {"name": "sessionId", "columnType": "string", "dataType": "string"},
    {"name": "hour", "columnType": "integer", "dataType": "time"},
    {"name": "views", "columnType": "integer", "dataType": "integer"},
    {"name": "clicks", "columnType": "integer", "dataType": "integer"}
  ],
  "rows": [
    ["15121011211.aac",0,5,1], ["15121011211.aac",1,3,0],
    ["15131113212.bfk",0,8,2], ["15131113212.bfk",13,2,0], …
  ],
  … // other properties
}
```

Fig. 3. Google analytics response data sample

Assume we are running an enterprise EL process using the implementation based on the proposed serverless approach. The EL process has the following constraints: (1) the process is run on a daily schedule, (2) Google Analytics is used as a source system,

(3) the data is represented by 100 different trackers containing 10,000,000 rows of data on average.

This results in 100,000 requests for extracting all of the data. With respect of rate limits the total execution time is estimated to be 100,000 requests/10 requests per second = 10,000 s. Consider this as effective execution time.

The traditional non-serverless approach involves a management node which continuously runs a scheduler. A virtual machine can play a role of a management node. In case of AWS the service which provides virtual machines is Elastic Cloud Compute (EC2). To reduce the cloud infrastructure costs the management node can also perform the function of a worker node. EC2 service provides a1.medium instances as a minimal general purpose virtual machine. Running a1.medium virtual machine in Ohio region for 30 days results in $0.0255 per hour × 24 h per day × 30 days = $18.36. The time-based resources utilization is measured as effective execution time divided by real execution time: 10,000 s/24 h = 11.6%.

The prototype of the proposed serverless approach has been implemented in form of a container run by a serverless container management AWS service: AWS Batch. AWS Batch does not require any additional payments but the ones spent by the underlying resources of EC2. This means that running the prototype on the same infrastructure has the same price for a1.medium virtual machine: $0.0255 per hour.

Execution of the described enterprise EL scenario using the AWS-based implementation of the proposed approach has taken 3 h and 15 min = 11,700 s. Extra time is explainable by setting up the resources and requesting the data source on the tasks metadata. After the data extraction has finished the resources have been shut down till the next execution. Extrapolating the execution evaluation to 30 days the resulting cost is estimated as $0.0255 per hour × 11,700 s per execution × 1 execution per day × 30 days = $2.49. The utilization of the prototype implementation equals to 10,000 s / 11,700 s = 85.5%.

Using the proposed approach reduces the infrastructure costs by 86.4% and increases the resources utilization by 85.5 − 11.6 = 73.9%.

6 Discussion and Conclusions

In the article we proposed and approach to deal with the batch Extract-Load processes for cloud Data Lakes and described an architecture of a serverless batch Extract-Load system. The proposed approach fits for a use-case of discontinuous batch jobs run in a serverless execution environment. Single job execution is dedicated into data extraction, partitioning and deduplication stages. After the data is stored into Data Lake persistent storage area it can be further processed by the outer ELT infrastructure.

The Amazon Web Services-based prototype implementing the architecture is evaluated according to a proposed evaluation method. The evaluation demonstrates that using the proposed approach significantly reduces cloud infrastructure costs and increases the resources utilization.

The described architecture is not limited to Amazon Web Services as the only supported cloud provider. The used services have analogues in other clouds [35], please see Table 2.

Table 2. Cloud providers services correspondence

Purpose	AWS	Azure	GCP
Storage	S3	Data lake service	Cloud storage
Tasks definitions storage	Dynamo DB	Cosmos DB	Cloud datastore
Orchestration	Step functions	Data factory	Cloud composer
FaaS	Lambda	Functions	Cloud functions
Containers service	Batch	Batch	Cloud run
PubSub service	SQS	Queue storage	Cloud PubSub

One of the main discussion points is the serverless paradigm limitations.

First risk is vendor lock: all the serverless resources are fully managed and cannot be easily migrated to another cloud provider unless some cross-cloud framework is used.

Another limitation is unpredictable resources availability: shared execution environment may need to wait up to several minutes to find available execution slots to execute the requested job. When the resources are payed on-demand or even reserved then they are available all the time. But this requires resources maintenance and may cause underutilization.

The negative effect of serverless paradigm is a cold start effect. This is a typical behavior of FaaS services: the first run of a function requires time for environment startup. But in terms of long-running data application jobs this is a minor drawback.

Acknowledgments. The research is financially supported by the Russian Foundation for Basic Research, projects 18-07-01434, 18-29-22096. The research was carried out using infrastructure of shared research facilities CKP «Informatics» (http://www.frccsc.ru/ckp) of FRC CSC RAS.

References

1. Lenzerini, M.: Data integration: a theoretical perspective. In: Proceedings of the Twenty-First ACM SIGMOD-SIGACT-SIGART Symposium on Principles of Database Systems, Madison, Wisconsin, pp. 233–246 (2002). https://doi.org/10.1145/543613.543644
2. Calvanese, D., de Giacomo, G., Lenzerini, M., Nardi, D.: Data integration in data warehousing. Int. J. Cooper. Inf. Syst. **10**(3), 237–271 (2001)
3. Davenport, R.J.: ETL vs ELT. A subjective view. Commercial aspects of BI. Insource House (2008)
4. Marín-Ortega, P., Dmitriyev, V., Abilov, M., Gómez, J.: ELTA: new approach in designing business intelligence solutions in era of big data. Proc. Technol. **16**, 667–674 (2014). https://doi.org/10.1016/j.protcy.2014.10.015
5. Miloslavskaya, N., Tolstoy, A.: Big data, fast data and data lake concepts. Proc. Comput. Sci. **88**, 300–305 (2016). https://doi.org/10.1016/j.procs.2016.07.439
6. Khine, P., Wang, Zh.: Data lake: a new ideology in big data era. In: ITM Web Conference, vol. 17, p. 03025 (2018). https://doi.org/10.1051/itmconf/20181703025
7. Shepherd, A., et al.: Opportunities and challenges associated with implementing data lakes for enterprise decision-making. Issues Inf. Syst. **19**(1), 48–57 (2018)

8. Munshi, A., Mohamed, Y.: Data lake lambda architecture for smart grids big data analytics. IEEE Access **6**, 40463–40471 (2018). https://doi.org/10.1109/ACCESS.2018.2858256

9. Pandey, S., Karunamoorthy, D., Buyya, R.: Workflow engine for clouds. In: Cloud Computing: Principles and Paradigms, chap. 12 (2011). https://doi.org/10.1002/9780470940105.ch12

10. Malik, M.: Cloud computing-technologies. Int. J. Adv. Res. Comput. Sci. **9**, 379–384 (2018)

11. Gannon, D., Barga, R., Sundaresan, N.: Cloud-native applications. IEEE Cloud Comput. **4**(5), 16–21 (2017). https://doi.org/10.1109/MCC.2017.4250939

12. McGrath, G., Brenner, P.: Serverless computing: design, implementation, and performance. In: 2017 IEEE 37th International Conference on Distributed Computing Systems Workshops (ICDCSW), Atlanta, GA, pp. 405–410 (2017). https://doi.org/10.1109/ICDCSW.2017.36

13. Baldini, I., et al.: Serverless computing: current trends and open problems. In: Chaudhary, S., Somani, G., Buyya, R. (eds.) Research Advances in Cloud Computing, pp. 1–20. Springer, Singapore (2017). https://doi.org/10.1007/978-981-10-5026-8_1

14. Elgendy, N., Elragal, A.: Big data analytics: a literature review paper. In: Perner, P. (ed.) ICDM 2014. LNCS (LNAI), vol. 8557, pp. 214–227. Springer, Cham (2014). https://doi.org/10.1007/978-3-319-08976-8_16

15. Jovanovic, P., Romero, O., Abelló, A.: A unified view of data-intensive flows in business intelligence systems: a survey. In: Hameurlain, A., Küng, J., Wagner, R. (eds.) Transactions on Large-Scale Data- and Knowledge-Centered Systems XXIX. LNCS, vol. 10120, pp. 66–107. Springer, Heidelberg (2016). https://doi.org/10.1007/978-3-662-54037-4_3

16. Kim, Y., Lin, J.: Serverless data analytics with flint. In: 2018 IEEE 11th International Conference on Cloud Computing (CLOUD), San Francisco, CA, pp. 451–455 (2018). https://doi.org/10.1109/CLOUD.2018.00063

17. Gerber, A., le Roux, P., Kearney, C., van der Merwe, A.: The Zachman framework for enterprise architecture: an explanatory IS theory. In: Hattingh, M., Matthee, M., Smuts, H., Pappas, I., Dwivedi, Y.K., Mäntymäki, M. (eds.) I3E 2020. LNCS, vol. 12066, pp. 383–396. Springer, Cham (2020). https://doi.org/10.1007/978-3-030-44999-5_32

18. Zhou, G., Xie, Q., Hu, Y.: E-LT integration to heterogeneous data information for SMEs networking based on E-HUB. In: Fourth International Conference on Natural Computation, Jinan, pp. 212–216 (2008). https://doi.org/10.1109/ICNC.2008.77

19. Sabtu, A., et al.: The challenges of extract, transform and loading (ETL) system implementation for near real-time environment. In: 2017 International Conference on Research and Innovation in Information Systems (ICRIIS), Langkawi, pp. 1–5 (2017). https://doi.org/10.1109/ICRIIS.2017.8002467

20. Couvares, P., Kosar, T., Roy, A., Weber, J., Wenger, R.: Workflow in Condor. In: Taylor, I., Deelman, E., Gannon, D., Shields, M. (eds.) Workflows for e-Science. Springer, Heidelberg (2007)

21. Tannenbaum, T., Wright, D., Miller, K., Livny, M.: Condor: a distributed job scheduler. In: Beowulf Cluster Computing with Windows, pp. 307–350. MIT Press, Cambridge (2001)

22. Deelman, E., Blythe, J., Gil, Y., et al.: Mapping abstract complex workflows onto grid environments. J. Grid Comput. **1**, 25–39 (2003). https://doi.org/10.1023/A:1024000426962

23. Ludäscher, B., et al.: Scientific workflow management and the Kepler system. Concurr. Comput.: Pract. Exper. **18**, 1039–1065 (2006). https://doi.org/10.1002/cpe.994

24. Oinn, T., et al.: Taverna: a tool for the composition and enactment of bioinformatics workflows. Bioinformatics **20**(17), 3045–3054 (2004). https://doi.org/10.1093/bioinformatics/bth361

25. Jacob, J., et al.: Montage: a grid portal and software toolkit for science-grade astronomical image mosaicking. Int. J. Comput. Sci. Eng. (IJCSE) **4**(2) (2009). https://doi.org/10.1504/IJCSE.2009.026999

26. Abramovici, A., Althouse, W.: LIGO: The laser interferometer gravitational-wave observatory. Science **256**(5055), 325–333 (1992). https://doi.org/10.1126/science.256.5055.325

27. Deelman, E., et al.: Pegasus: mapping scientific workflows onto the grid. In: Dikaiakos, M.D. (ed.) AxGrids 2004. LNCS, vol. 3165, pp. 11–20. Springer, Heidelberg (2004). https://doi.org/10.1007/978-3-540-28642-4_2

28. Jiang, Q., Lee, Y.C., Zomaya, A.Y.: Serverless execution of scientific workflows. In: Maximilien, M., Vallecillo, A., Wang, J., Oriol, M. (eds.) ICSOC 2017. LNCS, vol. 10601, pp. 706–721. Springer, Cham (2017). https://doi.org/10.1007/978-3-319-69035-3_51

29. Alonso, G., Hagen, C., Agrawal, D., El Abbadi, A., Mohan, C.: Enhancing the fault tolerance of workflow management systems. IEEE Concurr. **8**(3), 74–81 (2000). https://doi.org/10.1109/4434.865896

30. Couvares, P., Kosar, T., Roy, A., Weber, J., Wenger, K.: Workflow management in condor. In: Taylor, I.J., Deelman, E., Gannon, D.B., Shields, M. (eds.) Workflows for e-Science, pp. 357–375. Springer, London (2007). https://doi.org/10.1007/978-1-84628-757-2_22

31. Sun, L., Franklin, M., Krishnan, S., Xin, R.: Fine-grained partitioning for aggressive data skipping. In: Proceedings of the 2014 ACM SIGMOD International Conference on Management of Data (SIGMOD 2014), pp. 1115–1126. Association for Computing Machinery, New York (2014). https://doi.org/10.1145/2588555.2610515

32. Liu, X., Iftikhar, N.: Ontology-based big dimension modeling in data warehouse schema design. In: Abramowicz, W. (ed.) BIS 2013. LNBIP, vol. 157, pp. 75–87. Springer, Heidelberg (2013). https://doi.org/10.1007/978-3-642-38366-3_7

33. Mandagere, N., Zhou, P., Smith, M., Uttamchandani, S.: Demystifying data deduplication. In: Proceedings of the ACM/IFIP/USENIX Middleware 2008 Conference Companion (Companion 2008), pp. 12–17. Association for Computing Machinery, New York (2008). https://doi.org/10.1145/1462735.1462739

34. Getoor, L., Machanavajjhala, A.: Entity resolution: theory, practice & open challenges. Proc. VLDB Endow. **5**(12), 2018–2019 (2012). https://doi.org/10.14778/2367502.2367564

35. Perez-artega P.A., Guzmán L., Denneulin Y.: Cost comparison of lambda architecture implementations for transportation analytics using public cloud software as a service. Spec. Session Softw. Eng. Serv. Cloud Comput. (2018). https://doi.org/10.5220/0006869308550862

Data Management in Semantic Web

Pragmatic Interoperability and Translation of Industrial Engineering Problems into Modelling and Simulation Solutions

Martin T. Horsch[1,2]($\boxtimes$)(iD), Silvia Chiacchiera[1](iD), Michael A. Seaton[1](iD),
Ilian T. Todorov[1](iD), Björn Schembera[2](iD), Peter Klein[3](iD),
and Natalia A. Konchakova[4](iD)

[1] STFC Daresbury Laboratory, UK Research and Innovation,
Keckwick Ln, Daresbury, Cheshire WA4 4AD, UK
{martin.horsch,silvia.chiacchiera,michael.seaton,
ilian.todorov}@stfc.ac.uk
[2] High Performance Computing Center Stuttgart,
Nobelstr. 19, 70569 Stuttgart, Germany
{martin.horsch,bjoern.schembera}@hlrs.de
[3] Fraunhofer Institute for Industrial Mathematics,
Fraunhofer-Platz 1, 67663 Kaiserslautern, Germany
peter.klein@itwm.fraunhofer.de
[4] Institute of Surface Science, Helmholtz-Zentrum Hereon,
Max-Planck-Str. 1, 21502 Geesthacht, Germany
natalia.konchakova@hereon.de

Abstract. Pragmatic interoperability between platforms and service-oriented architectures exists whenever there is an agreement on the roles of participants and components as well as minimum standards for good practice. In this work, it is argued that open platforms require pragmatic interoperability, complementing syntactic interoperability (*e.g.*, through common file formats), and semantic interoperability by ontologies that provide agreed definitions for entities and relations. For consistent data management and the provision of services in computational molecular engineering, community-governed agreements on pragmatics need to be established and formalized. For this purpose, if ontology-based semantic interoperability is already present, the same ontologies can be used. This is illustrated here by the role of the "translator" and procedural definitions for the process of "translation" in materials modelling, which refers to mapping industrial research and development problems onto solutions by modelling and simulation. For associated roles and processes, substantial previous standardization efforts have been carried out by the European Materials Modelling Council (EMMC ASBL). In the present work, the Materials Modelling Translation Ontology (MMTO) is introduced, and it is discussed how the MMTO can contribute to formalizing the pragmatic interoperability standards developed by EMMC ASBL.

© Springer Nature Switzerland AG 2021
A. Sychev et al. (Eds.): DAMDID/RCDL 2020, CCIS 1427, pp. 45–59, 2021.
https://doi.org/10.1007/978-3-030-81200-3_4

Keywords: Pragmatic interoperability · Applied ontology · Process data technology · European Materials Modelling Council

1   Introduction

The capabilities of service, software, and data architectures increase greatly if they are able to integrate a variety of heterogeneous resources into a common framework. In general, this requires an exchange of information with a multitude of systems of resources, each of which follows the paradigm and structure, or *language*, favoured by its designers. As the number n of such systems increases, establishing and maintaining a direct $1 : 1$ *compatibility* between each pair of standards (*e.g.*, file formats) becomes impractical, considering that $n(n-1)$ converters would need to be developed and updated as each of the relevant standards is modified; beside the unfavorable scaling of the effort required to develop such an architecture, this would also presuppose that the designers of each system understand all other systems and are interested in ensuring a compatibility with each of them, neither of which can be taken for granted. Instead, $n : 1 : n$ *interoperability* based on a single intermediate standard only requires $2n$ mappings; if the interoperability standard has the approval of a significant community, which can be expected whenever the number of participating systems is large enough, developers have an intrinsic interest in maintaining the interoperability. For this purpose, they merely need to keep track of changes to their own system and the common intermediate standard.

Hence, interoperability is generally the favoured approach to integrating distributed and heterogeneous infrastructures. Since this addresses a problem of languages, the aspects of interoperability can be categorized according to their relation to three major areas of the theory of formal languages: Syntax, semantics, and pragmatics – or how to write correctly (according to a given format or grammar), how to associate a meaning with the communicated content (by which data items become information), and *how to deal with information* and transactions that involve an exchange of information. While well-known and well-established approaches for ensuring syntactic and semantic interoperability exist, *pragmatic interoperability* has not acquired the same degree of attention.

Software and data architectures often neglect to explicitly formulate any requirements at the level of pragmatics, since they are assumed to be guaranteed by institutional procedures (*e.g.*, who is given an account, and who may ingest data). However, this delegation of responsibilities cannot be upheld for *open infrastructures* where anybody is invited to participate and to which a multitude of external tools and platforms connect, each of which may have its own users, roles, service definitions, access regulations, interfaces, and protocols. Accordingly, finding that semantic interoperability cannot reach its goals if it is not supplemented by an agreement on "what kind of socio-technical infrastructure is required," it has been proposed to work toward a universal *pragmatic web* [38]; in full consequence, this would add a third layer to the world-wide web infrastructure, operating on top of the semantic web and hypertext/syntactic web layers.

This raises the issue of requirements engineering (*i.e.*, specifying and implementing requirements) for service-oriented infrastructures, which becomes non-trivial whenever "stakeholders do not deliberately know what is needed" [42]. Previous work has established that ontologies are not only a viable tool for semantic interoperability, but also for enriching the structure provided for the semantic space by definitions of entities, relations, and rules that are employed to specify jointly agreed pragmatics [38,41]; to provide additional procedural information, *workflow patterns* have been suggested as a tool [40], *e.g.*, employing the Business Process Model and Notation (BPMN) [1]. Since BPMN workflow diagrams can be transformed to RDF triples on the basis of an ontology [34], this approach is well suitable for domains of knowledge where ontologies already exist.

The present work follows a similar approach; it intends to contribute to the aim of the European Materials Modelling Council (EMMC ASBL) to make services, platforms, and tools for modelling and simulation of fluid and solid materials interoperable at all levels, which includes pragmatic interoperability. The workflow pattern standard endorsed and developed by EMMC ASBL is MODA (*i.e.*, Model Data) [7], which as an ontology becomes OSMO, the ontology for simulation, modelling, and optimization; this ontology development [21] constitutes the point of departure for the present discussion. One of the concepts at the core of this line of work is that of materials modelling *translation*, *i.e.*, the process of guiding an industrial challenge toward a solution with the help of modelling [13,25,32]. The experts that facilitate this process are referred to as *translators*; they provide a service for companies and can be either academics, software owners, internal employees of a company, or independent engineers [26]. Previous work on data science pragmatics by Neff *et al.* [29] concludes that it is particularly relevant to "get involved in observing the day-to-day practices of the work of data science" when addressing a scenario that "requires translation across multiple knowledge domains" to "make data valuable, meaningful, and actionable." This is the case here as well.

The remainder of this work is structured as follows: Sect. 2 introduces the approach to interoperability established by EMMC ASBL (and associated projects), the relevant definitions of roles and best practices concerning materials modelling translation, and how ontologies can be employed in this context. For this purpose, Sect. 3 introduces the main contribution from the present work, the Materials Modelling Translation Ontology (MMTO) version 1.3.7, together with OSMO version 1.7.5 which is extended in comparison to the previously published version 1.2 [21]. Section 4 discusses the identification of key performance indicators. Finally, a conclusion is given in Sect. 5.

2   Interoperability in Materials Modelling

2.1   Review of Materials Modelling, MODA, and Ontologies

Where a physically based modelling approach is followed, physical equations are employed jointly with materials relations that parameterize and complement the physical equations, *e.g.*, for a particular substance. The combination of physical equations and materials relations is referred to as the system of governing equations; on the basis of the Review of Materials Modelling [9], common physical equation types are identified categorized into four groups according to their granularity level: Electronic, atomistic, mesoscopic, or continuum. Subsequent to the review activity and the agreement on a basic nomenclature as formalized by the Review of Materials Modelling [9], the EMMC coordination and support action developed MODA, a semi-formalized simplified graph representation for simulation workflows [7]; this notation, which is immediately intelligible to human readers, but not immediately machine-processable, was further extended to permit the inclusion of graph elements that represent logical data transfer [21]. In MODA graphs, there are four classes of vertices, which are here referred to as *sections*:

1. Use case, *i.e.*, the physical system to be simulated.
2. Materials model, *i.e.*, the system of governing equations, with one or multiple physical equations and materials relations.
3. Solver, *i.e.*, the numerical solution of the model in terms of the variables that occur in the governing equations explicitly (and nothing beyond this).
4. Processor, *i.e.*, any computational operation beyond the above.

For each section, the MODA standard contains a list of text fields, which are here referred to as *aspects*, where more detailed information can be provided; however, since this is plain text, it is not immediately possible to extract semantically annotated content from this representation automatically. In logical data transfer graphs, additionally, there are vertices for *logical resources* that store *logical variables*, *i.e.*, abstractions of quantities and data structures that are exchanged between sections; the representation of the workflows and the flow of information is conceptual, or *logical*, in the sense that it does not carry any information on how the exchange of data is realized technically [21].

Semantic technology, centered on the use of ontologies as a tool, is increasingly applied to data management in all areas, including computational chemistry and molecular engineering [16,39]; beside ontologies, other types of semantic assets, such as XML schemas, can be employed [37]. As a feature of the semantic web, semantic assets can link to entity definitions from external sources, facilitating distributed development and complex multi-tier architectures. The highest level of abstraction is usually given by a top-level ontology. These components of the semantic web, such as the Basic Formal Ontology [2], DOLCE [6], ThingFO [30], and the Unified Foundational Ontology [19], are largely domain-independent and reused frequently in very diverse contexts; at the top level, philosopical concerns are at least as significant as the practical demands of research data technology. EMMC ASBL advances its own top-level ontology, the European Materials and Modelling Ontology (EMMO), *cf.* Francisco Morgado *et al.* [17], which implements the ontological paradigms of mereosemiotics [23], combining mereotopology [3] with semiotics following Peirce [31]. To achieve interoperability within the framework of the projects and infrastructures involving the EMMC community, all lower-level, *i.e.*, domain-specific ontologies need to be aligned with the EMMO.

In particular, for the Virtual Materials Marketplace project, which develops a platform where services and solutions related to computational molecular engineering can be traded, and with which multiple other platforms are expected to interoperate, it is essential to standardize the semantic space, *e.g.*, for the exchange of information during data ingest and data retrieval [24]; this includes the characterization of services, models, documents, data access, *etc.*, and may involve communication with external resources such as model and property databases. The domain ontologies that are developed by the Virtual Materials Marketplace project for this purpose are referred to as *marketplace-level ontologies* [22,24]; the marketplace-level ontology OSMO is directly based on the MODA workflow graph standard as well as its logical data transfer extension [21]. Thereby, a section from MODA, *e.g.*, a solver, becomes a `osmo:section` entity, *e.g.*, a `osmo:solver`. However, in MODA, the aspects (entries) of a section can only contain plain text; by using the relation `osmo:has_object_content` from OSMO, it becomes possible to point to semantically characterized content defined anywhere on the semantic web, including individuals and classes from OSMO and other ontologies. Similarly, OSMO formalizes the workflow graph elements and the exchanged logical variables. In this way, the MODA standard becomes machine-processable through OSMO. The materials modelling translation ontology from the present work, *cf.* Sect. 3, is based on this approach; it is closely connected to OSMO, and by extending the structure of the accessible semantic space to additionally deal with translation in materials modelling, it explicitly builds on OSMO and implicitly generalizes the MODA standard.

2.2   Specification of Roles and Processes

The role of the materials modelling *translator* is defined in the EMMC Translators' Guide [25]; a translator needs to be able to bridge the "language gap" between industrial end users, software owners, and model providers who are usually academics. The work of a translator aims at delivering not just modelling results, but a valuable and beneficial solution for a problem from industrial engineering practice [26]. Previous work by Faheem *et al.* [15] has shown that ontology engineering can support the implementation of a "mapping scheme from the problem domain to the computer domain" with a focus on deployment on high performance computing architectures, which as a challenge is roughly of the same type as translation in materials modelling – in particular, considering potential solutions such as the *translation router* developed by the Virtual Materials Marketplace project [24]. An instance of the translation process [32], some agreed features of which are codified by the EMMC Translators' Guide [25] and the EMMC TC Template [13], is referred to as a *translation case* (TC). It begins with exploring and understanding the *business case* and the *industrial case*, or multiple relevant business cases and/or industrial cases, which describe the socioeconomic objectives and boundary conditions, *cf.* Sect. 3.

Role definitions are known to be helpful in establishing sustainable good practices in data stewardship; *e.g.*, this is illustrated by recent work proposing the position of the *Scientific Data Officer* (SDO), jointly with providing a role definition that is tailor-made for addressing major concerns from research data management in high performance computing [36]. The responsibilities associated with this role relate to technical, organizational, ethical, and legal aspects of data stewardship [28]. One of the most important tasks that an SDO is expected to perform is data annotation, reducing or eliminating dark data [35]; since data can only be curated with the help of metadata, concrete tasks include adaptation of existing metadata models to a use case and the support of automated metadata extraction. Moreover, the SDO's responsibilites also include a mediation role between different groups of interest, *e.g.*, between scientists and the operators of computing and storage facilities. This mediation role has a high impact for pragmatic interoperability, since many problems arise when different technical languages, terminologies or jargons are conflicting. In this way, the SDO is in a position that shares certain characteristics with that of the materials modelling translator, particularly if metadata are seen as a form of communication as proposed by Edwards *et al.* [11].

Translation can be a process with multiple iterations. Thereby, the active and regular contact with the end user (*i.e.*, the industrial client of the translator) is a prerequisite for an effective and successful working relationship: The translator needs to be in communication with the client during the whole project duration to discuss regularly the project dynamics, possible changes to the line of work and development, and any other relevant feedback. The level of detail required for a modelling and simulation based contribution to an economic analysis of a value-added chain makes it necessary to go beyond computational molecular engineering in the strict sense, since eventually, key performance indicators of processes and products need to be optimized. The subsequent sections propose a solution for documenting these processes, providing the required business case, industrial case, and translation case descriptions (Sect. 3) as well as key performance indicators (Sect. 4).

3 Materials Modelling Translation Ontology

The most recent updates of the MMTO (presently,[1] version 1.3.7) and OSMO (presently,[2] version 1.7.5) are available as free software under the terms and conditions of the GNU Lesser General Public License version 3. Compared with the previously published version of OSMO [21], its present version generalizes the section structure from MODA by introducing `osmo:application_case` as a new superclass of `osmo:use_case`. Beside use cases, in this way, business cases (`mmto:business_case`), industrial cases (`mmto:industrial_case`), and translation cases (`mmto:translation_case`) become subclasses of `osmo:application_case` and can be dealt with in a similar way as the sections from MODA. The TC *aspects*, *cf.* Table 1, directly correspond to the text fields from the EMMC

---

[1] URL: http://www.molmod.info/semantics/mmto.ttl, as of 11th May 2021.
[2] URL: http://www.molmod.info/semantics/osmo.ttl, as of 11th May 2021.

TC Template [13], except that the MMTO permits the provision of semantically characterized content; this follows the approach from OSMO, which delivers the same feature for the text fields from MODA. The aspects by which business cases and industrial cases are described in the MMTO are given in Tables 2 and 3. Thereby, a business case can represent any economic problem at the management level, whereas an industrial case refers to an industrial engineering problem or an optimization problem in research and development. Within the translation process, a suitable approach based on modelling and simulation is identified and carried out; subsequently, the outcome is translated back to support an actionable decision at the business case and industrial case levels. The stages of the translation process according to the EMMC Translators' Guide [25], together with the corresponding MMTO entities, are reported in Table 4. To show how this would actually be realized on a digital marketplace, an illustrative exchange of communications taking place during a translation process (ordered as a sequence in time from top to bottom) is depicted in Fig. 1, together with the class hierarchy of the relevant branch of the MMTO.

Table 1. Aspects of a translation case (TC), `mmto:translation_case`, specified by the MMTO on the basis of the EMMC TC template [13].

aspect class name	content description
`mmto:tca_translator`	*translator(s) involved in the TC* content type: `evmpo:translator` [22]
`mmto:tca_end_user`	*involved end user(s), i.e., client(s) of the translator* content type: `evmpo:end_user` [22]
`mmto:tca_industrial_case`	*industrial case(s) associated with the TC* content type: `mmto:industrial_case`
`mmto:tca_business_case`	*business case(s) associated with the TC* content type: `mmto:business_case`
`mmto:tca_expected_outcome`	*expected outcome of the translation process* content type: plain text, *i.e.*, `xs:string`
`mmto:tca_pe_type`	*physical equation type(s) employed for modelling* content type: `osmo:physical_equation_type` [21]
`mmto:tca_discussion`	*summary of discussions with the end user* content type: plain text, *i.e.*, `xs:string`
`mmto:tca_kpi_model`	*employed KPI model(s)* content type: `mmto:kpi_model`
`mmto:tca_evaluation`	*evaluation (assessment) of the TC* content type: `vivo:translation_assessment` [22]
`mmto:tca_impact`	*impact and benefit to the end user; how does the* *TC contribute to improving processes/products?* content type: plain text, *i.e.*, `xs:string`
`mmto:tca_decision_support`	*employed decision support system(s)* content type: `osmo:decision_support_system`

Table 2. Aspects of a business case, `mmto:business_case`, in the MMTO.

aspect class name	content description
`mmto:bca_description`	*abstract or rough description* content type: plain text, *i.e.*, `xs:string`
`mmto:bca_industrial_case`	*related industrial case(s)* content type: `mmto:industrial_case`
`mmto:bca_red_zone`	*red zone(s), i.e., operational constraint(s)* content type: plain text, *i.e.*, `xs:string`
`mmto:bca_context`	*business-case context; revenue streams,* *risk management, distribution channels, etc.* content type: plain text, *i.e.*, `xs:string`
`mmto:bca_currency`	*budgeting currency* content type: `cao:Currency` from CAO [5]
`mmto:bca_contribution_to_cost`	*contribution to cost (in budgeting currency)* description type: plain text, *i.e.*, `xs:string` magnitude type: decimal, *i.e.*, `xs:decimal`
`mmto:bca_total_cost`	*total cost (in budgeting currency)* content type: decimal, *i.e.*, `xs:decimal`
`mmto:bca_contribution_to_benefit`	*contribution to benefit (in budgeting currency)* description type: plain text, *i.e.*, `xs:string` magnitude type: decimal, *i.e.*, `xs:decimal`
`mmto:bca_total_benefit`	*total benefit (in budgeting currency)* content type: decimal, *i.e.*, `xs:decimal`
`mmto:bca_return_on_investment`	*return on investment* content type: decimal, *i.e.*, `xs:decimal`
`mmto:bca_decision_support`	*employed decision support system(s)* content type: `osmo:decision_support_system`

The MMTO and OSMO are connected to the EMMO through the European Virtual Marketplace Ontology (EVMPO), a module which is developed jointly by the Virtual Materials Marketplace and MarketPlace projects, and the EMMO-VIMMP Integration component for ontology alignment [22,24]. Additionally, the MMTO employs the ISO 4217 standard for currency descriptions through the Currency Amount Ontology (CAO) module of the Financial Industry Business Ontology [5], *cf.* Table 2; it also refers to entity definitions from two further marketplace-level ontologies from the Virtual Materials Marketplace project [22,24]: Classes of agents (*e.g.*, `evmpo:end_user` and `evmpo:agent`) and messages (*e.g.*, `evmpo:interlocution` and `evmpo:statement`) from the EVMPO in combination with the VIMMP Communication Ontology, and the description of marketplace-interaction evaluations (here, `vivo:translation_assessment`) from the VIMMP Validation Ontology (VIVO), *cf.* Table 1.

Table 3. Aspects of an industrial case, `mmto:industrial_case`, in the MMTO.

aspect class name	content description
`mmto:ica_constraints`	*constraints: production capacity, supply chains, etc.*
	content type: plain text, *i.e.*, `xs:string`
`mmto:ica_data_source`	*data sources (documents, citations, databases, etc.)*
	content type: `iao:information_content_entity` [8]
`mmto:ica_data_access`	*data access conditions (licensing, ownership, etc.)*
	content type (description): plain text, *i.e.*, `xs:string`
`mmto:ica_decision_support`	employed decision support system(s)
	content type: `osmo:decision_support_system`

4 Key Performance Indicators

Key performance indicators (KPIs) can be used as a vehicle to map industrial problems onto modelling and simulation workflows [10]. In business administration and management, a KPI is understood to be a natural-language description of something which is a selling argument. This reflects the point of view corresponding to organizational roles that are comparably distant from research and development, *e.g.*, in sales or high-level management. In scenarios that arise in the context of such organizational roles, it necessarily appears to be most crucial to address concerns that are immediately relevant to business-to-administration, business-to-business, and business-to-customer relations [4].

We propose to reserve the keyword KPI (`mmto:key_performance_in-dicator`) to indicators (scalar quantities) that are directly relevant to characterizing, modelling, or optimizing such scenarios. On this basis, from the point of view of a materials modelling translator, two major distinctions need to be made:

1. Some KPIs are closely related to human sentience (aesthetics, haptics, taste, *etc.*). Studies aiming at gaining information on these quantities typically rely on market research and other empirical methods that involve human subjects; such indicators are referred to as *subjective* KPIs (`mmto:subjective_kpi`). Obversely, an *objective* KPI (`mmto:objective_kpi`) can be determined by a standardized process, *e.g.*, a measurement, experiment, or simulation, the result of which (assuming that it is conducted correctly) does not depend on the person that carries it out.
2. An objective KPI is *technological* (`mmto:technological_kpi`) if it is observed or measured within a technical or experimental process, referring directly to properties of the real product or manufacturing process; properties of a model, which are determined by simulation, are *computational* KPIs (`mmto:computational_kpi`).

The distinction between subjective and objective KPIs is similar to that between critical-to-customer and critical-to-quality measures [14,27,33]. The formulation given above, however, is more closely related to concepts from the

Table 4. MMTO representation of the stages of a materials modelling translation process (subclasses of `mmto:translation_step`), as specified by the EMMC Translators' Guide [25]. The numbers in the first column, which follow the EMMC Translators' Guide, are related to the stages by the datatype property `mmto:has_emmc_guide_no`. The considered sections, *i.e.*, individuals of the classes given in the third column without an asterisk, are related to the stages by `mmto:considers_section`. *Remark on no. 4: The relations connecting `mmto:translation_step_modelling` to `osmo:workflow_graph` is `mmto:considers_workflow`; **remark on no. 6: The relations connecting `mmto:translation_step_decision` to `osmo:decision_support_system` and `mmto:kpi_model`, respectively, are `mmto:has_step_decision_support`, and `mmto:considers_kpi_model`.

no.	MMTO class identifier and description	entities connected by relations
1	`mmto:translation_step_bc` *good understanding of the business case*	`mmto:business_case`
2	`mmto:translation_step_ic` *good understanding of the industrial case*	`mmto:industrial_case`
3	`mmto:translation_step_data` *analysis of the data from experiment and simulation available to the end user*	`osmo:use_case`
4*	`mmto:translation_step_modelling` *translation to simulation workflows*	`osmo:materials_model` and `osmo:workflow_graph`*
5	`mmto:translation_step_execution` *execution and validation strategy*	–
6**	`mmto:translation_step_decision` *evaluation of the simulation results to facilitate an actionable decision*	`osmo:decision_support_system`** and `mmto:kpi_model`**

EMMO. Due to its foundation on Peircean semiotics [31], it is straightforward in the EMMO to categorize signs by the way in which their interpretation depends on the subjective impression of an interpreter or observer [17,18,23]. In particular, the same distinction is made in EMMO version 1.0.0 beta [12]. Accordingly, this approach is best amenable to a prospective alignment of the MMTO with the EMMO and the approach to interoperability guided by EMMC ASBL.

The relation between properties accessible to materials modelling and the technological KPIs that are most immediately relevant to real industrial processes and products is necessarily indirect; it requires the mediation through a translation process and a TC as formalized above, which includes modelling KPIs as a function of other quantities, *i.e.*, the creation of KPI models (`mmto:kpi_model`). For the present purpose, a KPI model is given by a condition, correlation, or other formalism containing a set of variables, which can – but need not – be KPIs or other indicators, by which one or multiple KPIs are represented (*i.e.*, here, predicted, correlated, or modelled). In this way, KPI models can represent observables: Mathematical operators that map logical variables to scalar quantities, by which, *e.g.*, computational KPIs determined as the outcome of a complex simulation workflow can be correlated, and a technological KPI can be estimated on the basis of computational KPIs.

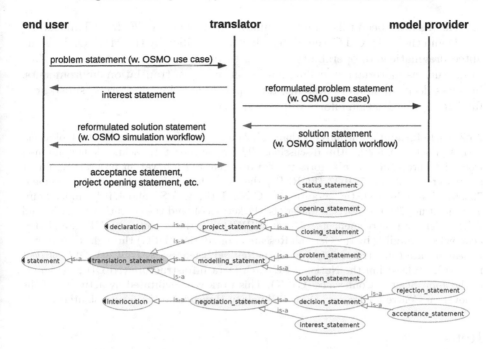

Fig. 1. Top: A possible interchange between participants of a materials modelling marketplace during steps 1 to 4 of a translation process, *cf.* Table 4, expressed in terms of `mmto:translation_statement` entities. Bottom: Diagram showing the transitive reduction of the `rdfs:subClassOf` relation (grey arrows) for the `mmto:translation_statement` branch of the MMTO, wherein `evmpo:declaration`, `evmpo:interlocution`, and `evmpo:statement` are entities from the European Virtual Marketplace Ontology (EVMPO) [22]; this diagram was generated using OWLViz [20].

5   Conclusion

By developing the MMTO, which extends the *section* concept from OSMO to cover business cases, industrial cases, and translation cases, a formalism was introduced by which translation in materials modelling can be represented in a way that implicitly also extends MODA, the pre-existing EMMC standard for simulation workflows. Just as OSMO is the ontology version of MODA, the MMTO is the ontology version of an implicit generalization of MODA by which, beside the simulation workflow itself, its socioeconomic context can be described. In this way, the MMTO is also a tool for representing the exchange of information during translation processes (*e.g.*, on KPIs) as a workflow, analogous to the formalization of MODA and logical data transfer graphs within OSMO. Since it is given as an ontology, the *aspects* from the MMTO (and from OSMO), which correspond to plain-text form entries in MODA, can contain links to entities defined elsewhere in the semantic web which can be immediately processed computationally, and to which automated reasoning can be applied. Where available, previous agreements have been taken into account in the form of the EMMC Translators' Guide and the EMMC TC Template, which are codified by the MMTO. To guarantee pragmatic interoperability between translation-related services and platforms such as materials modelling marketplaces, open translation environments, business decision support systems, and open innovation platforms, substantial further standardization efforts will be required.

Acknowledgment. The authors thank N. Adamovic, W. L. Cavalcanti, G. Goldbeck, and A. Hashibon for fruitful discussions. The co-author P.K. acknowledges funding from the European Union's Horizon 2020 research and innovation programme under grant agreement no. 721027 (FORCE), the co-author N.K. under grant agreement 723867 (EMMC-CSA), the co-authors S.C., M.T.H., M.A.S., and I.T.T. under grant agreement no. 760907 (Virtual Materials Marketplace), and the co-authors N.A.K. and P.K. under grant agreement no. 952903 (VIPCOAT); the co-authors M.T.H. and B.S. acknowledge funding by the German Research Foundation (DFG) through the National Research Data Infrastructure for Catalysis-Related Sciences (NFDI4Cat), DFG project no. 441926934, within the National Research Data Infrastructure (NFDI) programme of the Joint Science Conference (GWK). This work was facilitated by activities of the Innovation Centre for Process Data Technology (Inprodat e.V.), Kaiserslautern.

References

1. Allweyer, T.: BPMN 2.0: Introduction to the Standard for Business Process Modeling. BOD, Norderstedt, Germany, 2nd edn. (2016). ISBN 978-3-8370-9331-5
2. Arp, R., Smith, B., Spear, A.D.: Building Ontologies with Basic Formal Ontology. MIT Press, Cambridge, Massachusetts, USA (2015). ISBN 978-0-262-52781-1
3. Asher, N., Vieu, L.: Toward a geometry of common sense: A semantics and a complete axiomatization of mereotopology. In: Mellish, C. (ed.) Proceedings of the 14th IJCAI, pp. 846–852. Morgan Kaufmann, San Mateo, California, USA (1995). ISBN 978-1-55860-363-9
4. Barrientos, L.G., Sosa, E.R.C., García Castro, P.E.: Considerations of e-commerce within a globalizing context. Int. J. Manag. Inf. Syst. **16**(1), 101–110 (2012). https://doi.org/10.19030/ijmis.v16i1.6726
5. Bennett, M.: Reuse of semantics in business applications. In: Guizzardi, G., et al. (eds.) Ontologies in Conceptual Modeling and Information Systems Engineering (ONTO.COM/ODISE 2014). CEUR-WS, Aachen, Germany (2014)
6. Borgo, S., Masolo, C.: Ontological foundations of DOLCE. In: Poli, R., Healy, M., Kameas, A. (eds.) Theory and Applications of Ontology: Computer Applications, pp. 279–295. Springer, Dordrecht, Netherlands (2010). https://doi.org/10.1007/978-90-481-8847-5_13. ISBN 978-90-481-8846-8
7. CEN-CENELEC Management Centre: Materials modelling: Terminology, classification and metadata. CEN workshop agreement 17284, Brussels, Belgium (2018)
8. Ceusters, W.: An information artifact ontology perspective on data collections and associated representational artifacts. Stud. Health Technol. Inform. **180**, 68–72 (2012). https://doi.org/10.3233/978-1-61499-101-4-68
9. De Baas, A.F. (ed.): What makes a material function? Let me compute the ways. EU Publications Office, Luxembourg (2017). This document is by convention known as the Review of Materials Modelling (RoMM) (2017). ISBN 978-92-79-63185-6

10. Dykeman, D., Hashibon, A., Klein, P., Belouettar, S.: Guideline business decision support systems (BDSS) for materials modelling (2020). https://doi.org/10.5281/zenodo.4054009
11. Edwards, P.N., Mayernik, M.S., Batcheller, A.L., Bowker, G.C., Borgman, C.L.: Science friction: Data, metadata, and collaboration. Soc. Stud. Sci. **41**(5), 667–690 (2011). https://doi.org/10.1177/0306312711413314
12. EMMC ASBL: European Materials and Modelling Ontology, version 1.0.0 beta (2021). https://github.com/emmo-repo/ and https://emmc.info/emmo-info/. Accessed 19th April 2021
13. EMMC Coordination and Support Action: EMMC Translation Case Template (2017). https://emmc.info/emmc-translation-case-template/. Accessed 11th May 2021
14. EMMC Coordination and Support Action: Report on business related quality attributes for industry integration of materials modelling. Project deliverable report 6.3 (2018)
15. Faheem, H.M., König-Ries, B., Aslam, M.A., Aljohani, N.R., Katib, I.: Ontology design for solving computationally-intensive problems on heterogeneous architectures. Sustainability **10**, 441 (2018). https://doi.org/10.3390/su10020441
16. Farazi, F., et al.: OntoKin: an ontology for chemical kinetic reaction mechanisms. J. Chem. Inf. Model. **60**(1), 108–120 (2020). https://doi.org/10.1021/acs.jcim.9b00960
17. Morgado, J.F., et al.: Mechanical testing ontology for digital-twins: A roadmap based on EMMO. In: Castro, R.G., Davies, J., Antoniou, G., Fortuna, C. (eds.) SeDiT 2020: Semantic Digital Twins 2020, p. 3. CEUR-WS, Aachen (2020)
18. Goldbeck, G., Ghedini, E., Hashibon, A., Schmitz, G.J., Friis, J.: A reference language and ontology for materials modelling and interoperability. In: Proceedings of the NAFEMS World Congress 2019, p. NWC_19_86. NAFEMS, Knutsford, UK (2019)
19. Guizzardi, G., Wagner, G.: Using the Unified Foundational Ontology (UFO) as a foundation for general conceptual modeling languages. In: Poli, R., Healy, M., Kameas, A. (eds.) Theory and Applications of Ontology: Computer Applications, pp. 175–196. Springer, Dordrecht, Netherlands (2010). https://doi.org/10.1007/978-90-481-8847-5_8. ISBN 978-90-481-8846-8
20. Horridge, M.: OWLViz (2010). https://protegewiki.stanford.edu/wiki/OWLViz. Accessed 11th May 2021
21. Horsch, M.T., et al.: Semantic interoperability and characterization of data provenance in computational molecular engineering. Fluid Phase Equilib. **65**(3), 1313–1329 (2020). https://doi.org/10.1021/acs.jced.9b00739
22. Horsch, M.T., Chiacchiera, S., Cavalcanti, W.L., Schembera, B.: Data Technology in Materials Modelling. Springer, Cham (2021). https://doi.org/10.1007/978-3-030-68597-3.
23. Horsch, M.T., Chiacchiera, S., Schembera, B., Seaton, M.A., Todorov, I.T.: Semantic interoperability based on the European Materials and Modelling Ontology and its ontological paradigm: Mereosemiotics. In: Chinesta, F., Abgrall, R., Allix, O., Kaliske, M. (eds.) Proceedings of 14th WCCM-ECCOMAS 2020. Scipedia, Barcelona (2021). https://doi.org/10.23967/wccm-eccomas.2020.297

24. Horsch, M.T., et al.: Ontologies for the Virtual Materials Marketplace. KI - Künstliche Intell. **34**(3), 423–428 (2020). https://doi.org/10.1007/s13218-020-00648-9
25. Hristova-Bogaerds, D., et al.: EMMC Translators' Guide. Technical report, EMMC-CSA (2019). https://doi.org/10.5281/zenodo.3552260
26. Klein, P., et al.: Translation in materials modelling: Process and progress. Technical report, Zenodo (2021). https://doi.org/10.5281/zenodo.4729917
27. Machač, J., Steiner, F., Tupa, J.: Product life cycle risk management. In: Oudoza, C.F. (ed.) Risk Management Treatise for Engineering Professionals, pp. 51–72. IntechOpen, London, UK (2018). ISBN 978-1-78984-600-3
28. Mons, B.: Data Stewardship for Open Science. CRC, Boca Raton, Florida, USA (2018). ISBN 978-1-4987-5317-3
29. Neff, G., Tanweer, A., Fiore-Gartland, B., Osburn, L.: Critique and contribute: a practice-based framework for improving critical data studies and data science. Big Data **5**(2), 85–97 (2017). https://doi.org/10.1089/big.2016.0050
30. Olsina, L.: Analyzing the usefulness of ThingFO as a foundational ontology for sciences. In: Pons, C., Frías, M., Anacleto, A. (eds.) XXI Simposio Argentino de Ingeniería de Software, pp. 172–191. SADIO, Buenos Aires, Argentina (2020)
31. Peirce, C.S.: Peirce on Signs: Writings on Semiotic. University of North Carolina Press, Chapel Hill, North Carolina, USA (1991). ISBN 978-0-80784342-0
32. Pezzotta, M., et al.: Report on translation case studies describing the gained experience. Technical report, EMMC ASBL (2021). https://doi.org/10.5281/zenodo.4457849
33. Pries, K.H., Quigley, J.M.: Reducing Process Costs with Lean, Six Sigma, and Value Engineering Techniques. CRC, Boca Raton, Florida, USA (2013). ISBN 978-1-4398-8725-7
34. Rospocher, M., Ghidini, C., Serafini, L.: An ontology for the business process modelling notation. In: Garbacz, P., Kutz, O. (eds.) Formal Ontology in Information Systems: Proceedings of the Eighth International Conference, pp. 133–146. IOS, Amsterdam, Netherlands (2014). ISBN 978-1-61499-437-4
35. Schembera, B.: Like a rainbow in the dark: metadata annotation for HPC applications in the age of dark data. J. Supercomput. (3), 1–21 (2021). https://doi.org/10.1007/s11227-020-03602-6
36. Schembera, B., Durán, J.M.: Dark data as the new challenge for big data science and the introduction of the scientific data officer. Philos. Technol. **33**(1), 93–115 (2019). https://doi.org/10.1007/s13347-019-00346-x
37. Schembera, B., Iglezakis, D.: EngMeta: metadata for computational engineering. Int. J. Metadata Semant. Ontol. **14**(1), 26 (2020). https://doi.org/10.1504/IJMSO.2020.107792
38. Schoop, M., de Moor, A., Dietz, J.: The pragmatic web: a manifesto. Comm. ACM **49**(5), 75–76 (2006). https://doi.org/10.1145/1125979
39. Todor, A., Paschke, A., Heineke, S.: ChemCloud: chemical e-science information cloud. Nat. Prec. (2011). https://doi.org/10.1038/npre.2011.5642.1
40. Weber, J.H., Kuziemsky, C.: Pragmatic interoperability for eHealth systems: The fallback workflow patterns. In: 2019 IEEE/ACM 1st International Workshop on Software Engineering for Healthcare (SEH), pp. 29–36. IEEE, Piscataway, New Jersey, USA (2019). ISBN 978-1-72812252-6

41. Neiva, F.W., David, J.M.N., Braga, R., Campos, F.: Towards pragmatic interoperability to support collaboration: a systematic review and mapping of the literature. Informat. Softw. Technol. **72**, 137–150 (2016). https://doi.org/10.1016/j.infsof.2015.12.013
42. Wiesner, S., Thoben, K.D.: Requirements for models, methods and tools supporting servitisation of products in manufacturing service ecosystems. Int. J. Comput. Integr. Manuf. **30**(1), 191–201 (2016). https://doi.org/10.1080/0951192X.2015.1130243

Analysis of the Semantic Distance of Words in the RuWordNet Thesaurus

Liliya Usmanova[1]($\boxtimes$) (iD), Irina Erofeeva[1] (iD), Valery Solovyev[1] (iD),
and Vladimir Bochkarev[2] (iD)

[1] Institute of Philology and Intercultural Communication, Kazan Federal University, 18 Kremlyovskaya Street, 420008 Kazan, Russian Federation
[2] Institute of Physics, Kazan Federal University, 18 Kremlyovskaya Street, 420008 Kazan, Russian Federation

Abstract. The article presents the analyses of analogues in the RuWordNet thesaurus regarding their semantic distance. Consideration of analogues is by nature a comprehensive analysis for identifying of possible routes for each pair of words with application of both traditional linguistic methods and approaches of cognitive and corpus linguistics. The lack of semantic connections between some members of the series, as well as the inconsistency of their routes with the general linguistic representations established in the dictionaries serves the basis for distance spacing between words in the RuWordNet thesaurus. Methods for bringing analogues, which are remote in the RuWordNet thesaurus, closer together, were proposed basing on a comparative analysis of the data in the explanatory dictionaries and in the New Explanatory Dictionary of Synonyms of the Russian Language, taking into account semantic relations of inclusion or meanings intersection. Basing on the obtained data, recommendations to reduce the distances between analogues were formulated for RuWordNet; and general principles of analysis were proposed, which could be useful for RuWordNet verifying.

Keywords: Quasi-synonymy · Analogues · Thesaurus · RuWordNet · Synset · Synonymy · Semantic network

1 Introduction

The issue of distinguishing between synonyms and non-synonyms remains undecided in linguistics, and is solved at the level of the conceptual layer of semantics, depending on the dominant scientific paradigm. Synonymy within the framework of system centrism is considered a hierarchical combination of words of the same part of speech basing on the identity or similarity of their lexical meanings with clear boundaries between synonymic series, which enable to differentiate one synonymic series from another. In line with the cognitive approach, synonyms are considered the field structures or fragments of a continuous and fundamentally incomplete synonymic and, taken more broadly, associative-verbal network.

WordNet thesaurus [1], one of the most popular computer resources for automatic word processing, is built on synonymous series in English. A WordNet-like RuWordNet

© Springer Nature Switzerland AG 2021
A. Sychev et al. (Eds.): DAMDID/RCDL 2020, CCIS 1427, pp. 60–73, 2021.
https://doi.org/10.1007/978-3-030-81200-3_5

thesaurus [2] has recently become available for the Russian language. In this article, we describe the method of verification of the RuWordNet thesaurus based on a set of pairs of words-analogues.

A new insight in synonymy in terms of cognitive linguistics enables to interpret this phenomenon as unclear variety of lexical units within a synonymic family with a "floating" prototype (the term introduced by L.A. Araeva), its members being connected on the basis of a family resemblance [3]. The boundaries of the synonymic family are permeable and mobile, with many names located on its periphery, poorly ordered by vague equonimic, hyponymic and partitive relationships, or inconsistently differentiated along the language ecosystems.

It is customary in linguistics to distinguish absolute synonyms (doublets), lexical synonyms and quasi-synonyms in accordance with the semantic distance parameter imposed on words of this kind. Whereas the words of the first group express absolute identity of meanings and contexts, then lexical synonyms and quasi-synonyms differ in terms of interpretation agreement. The criteria of synonymy, according to Yu.D.Apresyan, are the equality of lexical meaning (which is represented using an elementary semantic language), and the coincidence of semantic characteristics (in particular, their valency): "To recognize two words as lexical synonyms, it is necessary and sufficient, (1) that they have a completely identical interpretation, i.e., they are translated as the same expression of the semantic language, (2) that they have the same number of active semantic valency, (3) that they belong to the same (deep) part of speech" [4, p. 223].

Quasi-synonyms differ from exact synonyms in terms of the first feature (their interpretations have a larger general part – in terms of terminology, but do not completely coincide) and do not differ by the second and third features.

Quasi-synonyms are defined as indirect, approximate synonyms; their meanings can vary in conceptual content, speaker's attitude, collocations, etc. and change depending on the context [5, p. 307]. These are words, which have a common component of meaning, but cannot replace each other in sentences, unlike synonyms.

Alongside with quasi-synonyms, some researchers tend to identify the so-called analogues, their meaning substantially intersecting with the general meaning of a given series of synonyms, although not reaching the degree of closeness to it, which constitutes synonymy itself: "...the analogue refers to the same part of speech as a given lexeme and has a similar meaning, but semantically it is further from this lexical unit than its synonyms. The most common examples of analogues are cohyponyms..." [6, p. 20]. Hyperonyms and hyponyms, as well as lexical units associated with a given lexical unit with a less defined semantic connection can be also included among the analogues [6, pp. 23–24].

The existence of different variants of approaching and similar semantic modifications in speech involves the search for a formalized method of distinguishing between quasi-synonyms and analogues. According to Yu.D.Apresyan, unlike analogues, it is possible to neutralize the semantic differences between quasi-synonyms in a number of positions. Moreover, different types of quasi-synonyms manifest themselves differently regarding neutralization, depending on the internal structure: generic-specific differences (inclusion of meanings) are usually neutralized, neutralization of type-specific differences (intersection of meanings), which is quite possible theoretically, in practice turns out to be quite a rare phenomenon [4, p. 235].

In the study of semantic proximity of words in a WordNet thesaurus, the distance between words in a graph of semantic relationships is often used [7–9]. In a series of works [10, 11] the semantic proximity of words was calculated over a large corpus of texts using standard methods for vector representation of words semantics. Then, for semantically close words in the corpus, their degree of closeness was evaluated. If such pairs of words turned out to be distant, then this became the subject of analysis and, in some cases, made it possible to identify mistakes in RuWordNet.

Also, for RuWordNet, statistical parameters were calculated, such as the average path length between the thesaurus concepts, and compared with the values of these parameters for WordNet [12].

2   Methodology

The object of our study has become a comparative analysis of analogues from the New Explanatory Dictionary of Synonyms of the Russian Language (NEDS) [6] for the degree of their distance in the RuWordNet thesaurus [13]. The NEDS describes 2590 pairs of analogue words found manually by the authors. The distance between the words in the semantic connections graph – the length of the shortest path was chosen as a measure of closeness.

The following traditional linguistic methods, such as the descriptive method, the onomasiology method and the derivation analysis method were involved in the process of work, as well as the modern methods of cognitive and corpus linguistics.

A computer program was created to automate the process, which listed shortest routes between specified pairs of vertices for each pair of analogous words along the semantic relations in the thesaurus.

The program construed the network of semantic relations as an undirected graph, its vertices being the words and phrases, and the edges – the semantic relations connecting them. The degree of semantic similarity of the two words, as mentioned above, was estimated by the length of the shortest path in the graph connecting the vertices corresponding to these words. The shortest path length was found through the breadth-first search algorithm. Cormen et al. (2009) give its detailed description and pseudo-code [14]. An adjacency matrix was constructed at the first stage basing on the data of semantic relationships of the types of our interest extracted from RuWordNet (calculations were performed with and without respect to the domain relation). When creating the RuWordNet thesaurus, the following relations between synsets were taken into account: hyponym-hyperonym (genus-species), instance-class, antonymic relation, part-whole, reason, logical consequence, subject area (domain). Also, relations of part-of-speech synonymy, connecting the separated synsets, are defined between synsets belonging to different parts of speech, but expressing the same meaning. More details on the methodology for creating the RuWordNet thesaurus can be found in [5]. The adjacency matrix is 130.3k by 130.3k in size, and contains 4.32 million nonzero entries (2.94 million without domain). More details about the properties of the adjacency matrix can be found in [12].

Neighborhoods of increasing radius for a word A were sequentially constructed in order to find the length of the shortest path from a word A to a word B. The calculations stopped when, at a certain radius, the vertex B fell into the resulting neighborhood.

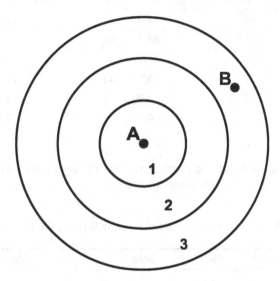

Fig. 1. Scheme for calculating the distance between words A and B

Figure 1 shows the scheme for calculating the distance between words A and B. Circles symbolically indicate the neighborhood of the word A, corresponding to a distance 1, 2 and 3, respectively.

The calculation results with the percentage ratio of grouped words are presented in the form of the Excel format Tables (Table 1, 2). Table 1 shows examples of analogues pairs with distances between them, calculated in two variants – with and without taking into account the relationship 'domain'.

Table 1. Distances between analogues-nouns in RuWordNet. Fragment.

Dominant	Analog	Distance	Except domain
water 'вода'	juice 'сок'	2	2
water 'вода'	dew 'роса'	2	2
water 'вода'	rain 'дождь'	3	4
driver 'водитель'	car enthusiast 'автолюбитель'	2	5
driver 'водитель'	chief 'шеф'	3	3
driver 'водитель'	Commander 'командир'	4	4
driver 'водитель'	owner-driver 'частник'	4	4
driver 'водитель'	working on the side 'левак'	4	5

Table 2. Distances between analogues-nouns in RuWordNet. Summary.

Distance	Relations		Relations except domain	
	Number of cases	%	Number of cases	%
1	68	24,73	68	24,73
2	95	34,55	65	23,64
3	58	21,09	65	23,64
4	42	15,27	47	17,09
5	11	4,00	24	8,73
6	1	0,36	6	2,18

Automatic calculations showed possible variants of target words connecting (in our case, the nouns) along all possible routes (Table 3).

Table 3. Paths between analogues-nouns in RuWordNet. Fragment.

Word	Step 1	Step 2	Step 3	Step 4	Step 5	Step 6
authori-ty'авторитет'	marked-man'известныйчеловек'	celebri-ty'знаменитость'	well-knownfig-ure'известность'	–	–	–
tramp 'бродяга'	waif 'беспризорник'	–	–	–	–	–
water 'вода'	liquid 'жидкость'	drink 'напиток'	–	–	–	–
imagina-tion'воображение'	notion 'измышление'	thinking 'мышление'	person 'персона'	soul 'душа'	–	–
pronunciation 'выговор'	talking 'говорение'	speech'выступление'	sermon 'проповедь'	edification 'назидание'	moral 'нравоучение'	–
fence 'забор'	wall 'стена'	house 'дом'	realty 'недвижимость'	plot 'делянка'	garden 'сад'	front garden 'палисадник'

The average number of possible shortest routes between these pairs is 63.7. Such a large number of routes is connected, apparently, with a significant number of synonyms in many synsets, which leads to a large number of parallel routes. It is clear that the largest number of shortest routes is characteristic of pairs of vertices that are more distant from each other.

Consider an example: paths in the thesaurus from the word magician to the word magician. The length of the shortest path between these words is 5. Group the words in different shortest paths for this pair, by distance from the original word in the graph:

A *wizard* → magical, sorcerous, man by field of activity (*кудеснический, чародейский, человекпосферамдеятельности*) → personal, man (adj.), human (adj.), athlete, sportsman, physical culturist, individual, individuum, avowant, personality, microcosm, person, persona, human being, human person (*личностный, людской, человеческий, человечий, атлет, спортсмен, физкультурник, индивид, индивидуум, лицо, личность, микрокосм, особа, персона, человек, человеческаяличность*) → artistic, performer, performer of composition, acrobat, woman-acrobat (*артистический, исполнитель, исполнительпроизведения, исполнительница, акробат, акробатка*) → magician (adj.), circus artist, circus actress, worker of circus, circus man, circus woman, circus performer, circus figure, circus worker, artist, actress (*фокуснический, артистцирка, артисткацирка, работникцирка, циркач, циркачка, цирковойартист, цирковаяартистка, цирковойдеятель, цирковойработник, артист, артистка*) → *a magician*.

The number of parallel shortest paths in this example is high due to the large number of synonyms at intermediate points of the word path.

Basing on the obtained data, recommendations were formulated for RuWordNet to reduce the distances between analogues, and general analysis principles were proposed, which can be useful for verifying RuWordNet.

Further work was carried out using the following methods. The conditions for the approaching of analogues in the structure of the thesaurus were identified applying *component and definitive analysis* for establishing the basic meanings of the analyzed words, for identifying differential semes, as well as the signs leading to the conceptual domain (DOMAIN).

The application of method of *lexical-semantic fields indicating* made it possible to identify the relationships between the elements of the field in a paradigmatic plan. The analysis procedure consisted in:

1. Consideration of the vocabulary definition of the target word as a detailed definition: identifier word + specifying words;
2. The field constituents identification was carried out further on the basis of their recurrence in the dictionaries of the same identifiers;
3. The analysis of the relationships between the elements of the field in a paradigmatic respect: synonymic, hyper-hyponymic relationships, cognates, part-of-speech synonymy, etc.

The cognitive interpretation of meanings was based on the techniques of *conceptual analysis of words*, which implies transition from the content of meanings to the content of concepts (considering etymologic data). Conclusions about the significance of certain cognitive traits in understanding the concept were drawn basing on the predominance either of the generic or the specific characteristics.

3 Results and Discussion

To date, the distinction between words that are similar in meaning is most clearly represented in the New Explanatory Dictionary of Russian Synonyms, which provides a detailed list of semantic features comparing elements of a synonymic series, indicating similarities and differences of synonyms, and also analyzes contextual (semantic, syntactic, compatible, morphological), pragmatic and other conditions for neutralizing the differences between synonyms. A group of analogues is also indicaed in each dictionary entry, being of interest for our study in terms of comparison with other dictionaries basing on the semantic principle, considering various kinds of connections and relations between words.

With this purpose, we turn to the RuWordNet digital thesaurus, which is known as a lexical resource in the field of automatic texts processing in Russian, created by transforming the Russian language thesaurus RuThes (RuTez) into the WordNet format developed at Princeton University of the United States [15]. Along with the synonymic relationships represented in synsets (sets of synonyms), the present thesaurus includes hyper-hyponymic relationships, as well as causation and succession relationships, domain relationships, word formation relationships, as well as relationships between word combinations and their components [5].

The semantic network of the thesaurus reflects the semantics of the domain in the form of concepts and relations, conditionally representing the form of a directed graph, the vertices of which correspond to the objects of the domain, and arcs (edges) define the relationships between them. This hierarchical principle of the thesaurus organization gives an idea of the vertical systematization of lexical sections, where everything is organized according to the laws of logical and semantic subordination, when a particular category falls into a general category or, conversely, a general category includes a particular category [16]. As M.V. Nikitin notes, "part-whole relationships" permeate the whole world from its top to the bottom, from micro- to macrocosm, from elementary particles to galaxies. They embrace things of all levels of complexity, organizing them into diverse multi-stage hierarchies of things, formed by the generality of signs and patterns inherent in it" [17, p. 387].

The cognitive orientation of this approach reflects two global dimensions of the semantic structure of the dictionary – hyponymic and partitive (partonymic), giving an idea of the unity and integrity of the world in its diversity, where everything is based on similarities/differences and interactions of things.

Due to the inclusion of these additional relationships between synsets, the RuWordNet thesaurus has a small world structure, which means a small diameter of semantic relations in the network, when you can reach from one point to another along the edges in a small number of steps.

Observations manifest that the considered pairs of analogues in terms of semantic relations are accessible in the structure of the thesaurus in a small number of steps. Loukachevitch (2019) notes that a distance of 4 or fewer steps is considered as small [10]. Out of the 2590 pairs of analogues obtained from NEDS by the continuous sampling method, 2548 words in RuWordNet are located within the distance of 4 or even fewer steps by the semantic relations, 39 – at a distance of 5, and 3 pairs – at a distance of 6 steps. The fact that 98% of the analogues are located at a close distance in the thesaurus is a serious argument in favor of a good organization of the structure of the thesaurus. The 4 steps as a measure of proximity are chosen so to say provisionally to a certain degree, based on the agreed intuition of the project executors. A long distance (5 steps or more) may be an indication of errors in the thesaurus or a consequence of an imperfect analytic algorithm.

As a result of the distances analysis between the analogue and the target word, RuWordNet has identified a number of assumptions regarding the causes of semantic shifts and an increase in the distance between them.

3.1 Skipping of Word Meanings

The most typical cases of the distance between analogues with different etiology of the phenomenon are given as examples.

1) A tramp – a hippie (бродяга – хиппи): The following chain in the thesaurus is built between these words: a tramp – a person in need of social protection (socially unprotected) – a resident of a retirement age – socio-demographic – youth – hippie (бродяга – лицо, нуждающееся в социальнойзащите (социально-незащищенный) – жительпенсионноговозраста – социально-демографический – молодежь – хиппи).

The lexical unit *tramp (бродяга)* is represented in two meanings in the RuWordNet thesaurus: Tramp 1 is interpreted as 'a person without a fixed place of residence', the concept going back to the hyperonym *a person who needs social protection (socially unprotected),* further –*a resident of a retirement age.* Tramp 2 (meaning 'vagabond, traveler') goes back to the hyperonym*traveler.* The word *hippie* goes back to hyperonyms 1. Youth environment; 2. Youth movement, i.e. representing only the external side of the phenomenon without indicating the object of its manifesting. Therefore, relying on interpretation given in the S.I. Ozhegov dictionary – 'a person, who has broken off with his environment, leading a vagabond lifestyle (usually together with others) generally people united in protest against existing relationships in society, challenging it with their passivity and inactivity' [18] – should indicate an additional meaning for the noun *hippie*– 'leading a vagabond lifestyle' and 'breaking with his environment'.

To reduce the distance between these nouns, it is also recommended to add the hyponym *hippie* in the synset Tramp 2 (this noun is not represented by any hyponyms), since explanatory dictionaries indicate common semes, which combine these two words.

2) A driver – working on the side (водитель – левак): The lexicographic portrait of the lexical unit *driver (водитель)* indicates its relevance to the persons operating ground transport.

In RuWordNet, the noun *working on the side (левак)* is represented in one meaning – 'left-wing politician' with the hyperonym*political figure (политдеятель),* further –*public figure (общественныйдеятель).* A domain *policy (политика)* indicates that this concept is related to a different semantic sphere than the word *driver (водитель).*

According to the Explanatory Dictionary by S.I. Ozhegov, in addition to the above named, there is also a meaning 2. – 'a worker illegally using his work time, tools or social work products for the personal profit'. Certain stylistic restrictions (colloquial style) with a negative emotional assessment are imposed on the use of the word.

Obviously, to bring the words in question closer together, a new meaning of the word *working on the side* should be added, linking it as a hyponym to the word *driver*.

3.2   Denotation to Different Semantic Areas

1) Imagination – dreaminess (воображение – мечтательность): The automatic calculation shows that the distance between these words in RuWordNet is 5 steps: imagination – invention – thinking – reflection – contemplation – dreaminess (воображение – измышление – мышление – раздумье – созерцательность – мечтательность).

The semantics of this polysemic lexeme in RuWordNet generally corresponds to the data in the explanatory dictionaries: Imagination 1 (invention, fiction) and Imagination 2 (mental ability). The first meaning goes back to the hyperonyms*lies, nonsense (враньё, ложь)*. In the second meaning, the noun has a hyperonym*mental ability (ментальнаяспособность)*. An analogue of dreaminess in Wiktionary is defined as 'inclination toward dreams, daydreaming'; in RuWordNet, it goes back to the hyperonym*contemplation (созерцательность)*. As you can see, *imagination* and *dreaminess* (воображение and мечтательность) belong to different semantic areas: ability and trait of character. To bring words closer, it is recommended to add a hyperonym*mental ability (ментальнаяспособность)* to synset dreaminess and include the cognate word *dream (мечта)*, which has a hyperonym*thought (мысль)*.

2) A driver – a car enthusiast (водитель – автолюбитель): A car enthusiast, according to Wiktionary, is an unprofessional car driver.

The word *driver* has 2 meanings in the RuWordNet thesaurus: Driver 1 (automobile transport driver) with the hyperonym*driver*, followed by *–the person in the field of activity (человекпосфередеятельности)*, and Driver 2 (transportation vehicle driver), with the hyperonym*–the person in the field of activity (человекпосфередеятельности)*, and further *–the individual (индивид)*. The noun car enthusiast has the meaning of 'car owner'. *The car enthusiast* goes back to the hyperonym*the owner (владелец)*, and further *–the property owner (владелецимущества)*.

Relation to various semantic areas (action (driving) –for the driver, state (ownership) –for the car enthusiast) increases the number of steps between words, despite the presence of common components in their meanings. In order to bring these concepts closer, it is proposed to add the meaning Car enthusiast 2 ('non-professional car driver') with the hyperonym*driver*. As a result, hyper-hyponymic relationships will be built between the considered concepts of *the driver* and *the car enthusiast*.

3) A wizard –a magician (волшебник – фокусник): An analogue to the word *wizard* in NEDS is a magician. Meanwhile, the distance between them in RuWordNet is calculated in 5 steps: a wizard –a person by field of activity –an individual –a performer –an actor –a magician (волшебник – человекпосфередеятельности – индивид

– исполнитель – артист – фокусник). An analysis of the hypo-hyperonymic connections presented in the synsets shows that the meaning of the noun *wizard* (волшебник) in RuWordNet as a whole coincides with the data of the explanatory dictionaries ('conjurer, sorcerer') ('колдун, чародей'), and has hyperonyms 1. *hero of a tale*; 2. *person by field of activity* (1. *герой сказки*; 2. *человек по сфере деятельности*).

A *magician* (фокусник), according to the dictionary, is an 'actor showing magic tricks'. The noun *magician* goes back in RuWordNet to the hyperonym *actor*, and further –*person of art*. As you can see, the wizard is associated with the fairy world, and the magician – with the real one. For making the distance closer, it is recommended that a noun *actor* shall be related as a hyponym for a synset *a person by field of activity*. Due to this, there will be short paths (2–3 steps) from *a wizard* to a *magician*.

4) Reprimand – moralizing (выговор – нравоучение): The distance between these words in RuWordNet is calculated in 5 steps: reprimand – enunciation– speech – sermon – precept – moralizing (выговор – выговаривание – выступление – проповедь – наказ – нравоучение).

In RuWordNet, the noun *reprimand* (выговор) is presented as polysemic: Reprimand 1 (disciplinary action), hyperonym 1. *Penal measure (мера наказания)*; 2. *Disciplinary action (дисциплинарное взыскание)*. Reprimand 2 (reprimand, enunciation) (выговор, произношение) hyperonym– *voice (глас)*. Reprimand 3 (to reprimand, make a reprimand) (выговаривать, делать выговор) hyperonym 1. *complaint (нарекание)*; 2. *blame (осуждение)* (further – *pronouncement (высказывание)*); 3. *irritation (раздражение)*.

The noun *moralizing* has the hyperonym *edification (назидание)*, further –*the recipe, advice (рецепт, совет)*. The long distance between the words is due to the attribution of hyperonyms to different semantic spheres.

Vocabulary definitions indicate the coincidence of the seme 'infusion' (внушение): according to the Explanatory Dictionary by S.I. Ozhegov, *reprimand* means 'strict verbal infusion; remark, which is a penalty, punishment for misconduct'. *Moralizing* has the meaning of 'teaching, suggesting moral rules'. The material in NEDS confirms our observations: we find an indication in the dictionary entry that *moralizing* is the closest to the word *infusion*, differing in a more abstract meaning and less close connection with a specific action of the addressee. Given that the *infusion* is a synonym for the word Reprimand 3, we recommend to include the lexical item *moralizing* in the composition of hyponyms for the noun Infusion 1 (*to reprimand, to make a reprimand*), especially that this noun lacks any represented hyponyms.

3.3 Skipping of Relationships

1) Din – hum (Гам – гудение): The calculations showed the distance between the words in 5 steps: din – ruckus – place – breakdown – pain – hum (гам – базар – местечко – надлом – боль – гудение). As can be seen from the presented chain, the increase in distance is due to the polysemantic words meaning mixing followed by semantic shift towards a different meaning. The long and winding path between the analogues in the thesaurus is conditioned by the fact that the connection between the related words *din* and *hum* (гам – гудение) is not taken into account. According to etymological data, both words go back to onomatopoeia (sound symbolism) *гу* (+ suffix –лъ) [19, p. 86].

Since din and hum (*гам* and *гул*) are hyponyms for the word *noise (шум)*, the inclusion of the word Hum 1 (to hum (to make a sound)) as a co-hyponym for the word *hum (гул)* would significantly reduce the distance between analogues.

*2) Smoke – exhaust (Дым – выхлоп):*The distance between the concepts comprises 6 steps: smoke – cloud – natural phenomenon – natural disaster – volcanic activity – removal – exhaust (дым – облако – природноеявление – природноебедствие – вулканическаяактивность – выведение – выхлоп). As you can see, the increase in the distance between analogues is related to the non-differentiation of the nature of this phenomenon origin: anthropogenic – 'lifting up gray clouds – volatile products of combustion' and of natural origin – haze fog, white steam, etc.'

The word smoke (дым) in the RuWordNet thesaurus goes back to the hyperonym*cloud (облако)*, further – *the set (совокупность)*. The concept of *exhaust (выхлоп)* is presented as polysemic: Exhaust 1 (engine exhaust) (hyperonym*harmful wastes* (вредныеотходы), further – *a harmful admixture* (вреднаяпримесь); Exhaust 2 (exhaust gas emissions) (hyperonym*atmospheric emissions* (атмосферныевыбросы), further –*anthropogenic pollution* (антропогенноезагрязнение).

According to the vocabulary interpretation, *exhaust (выхлоп)* is a derivative of the verb; *to exhaust:* 'of the engine: to release the exhaust gas'.

NEDS indicates a broader meaning of the word *smoke (дым)* as compared to the more specific concept of the word *exhaust*, which implies the presence of particles of soot, gasoline vapor, other volatile unburned residues and various gases. Unlike the word *smoke*, the words *exhaust (выхлопе)and exhaust gases* (выхлопныхгазах) are only applied in relation to the operation of an internal combustion engine, usually in a car. Thus, based on the above considerations regarding the relationship of inclusion between these concepts, we recommend adding *exhaust* to the hyponyms of the word *smoke*, which will significantly reduce the distance between these concepts.

*3) Fence– front garden (забор– палисадник):*According to calculations, the distance between these words in RuWordNet is measured in 6 steps: a fence – a wall – a house (building) – a residential house (residential building) – a single family house (manor type house) – a courtyard – a front garden (забор – стена – дом (здание) – жилойдом (жилойкорпус) – домдляоднойсемьи (домусадебноготипа) – подворье – палисадник).

Vocabulary definitions of lexemes *fence (забор)* ('fence, mostly wooden') and the *front garden (палисадник)* (a small fenced garden, flower bed in front of the house [18] and 'the fence, fencing, separating the local territory from the street' [20] indicate the common part of these words consisting in the idea of enclosing some part of the territory. We find confirmation of this in the NEDS, where the properties of these realities are revealed in detail and the method of creating these structures and the nature of the enclosed area are clarified. The authors of the dictionary divide two meanings of the word *front garden*: 1. *small fenced garden;* 2. (obsolete) *the front garden– 'short, light, permeant*. It encloses a small plot in front of the house where flowers are planted'. In RuWordNet, only the first value is taken into account.

A word *fence* in the RuWordNet thesaurus has two meanings: Fence 1 (Забор) (fence (fencing)), a hyperonym *poling (загородка)*, further – *construction (постройка)*; Intake 2 (Забор) (to take in (grab, grasp)). The word *front garden* goes back to hyperonyms 1. garden; 2. courtyard; 3. flower garden *(сад, подворье, цветник)*.

It seems the most natural to agree with the authors of the NEDS who recommend to bring these analogues closer together in RuWordNet by establishing hyper-hyponymic relations between them, putting the *front garden* in the status of a hyponym for Fence 1. Previously required to enter the value of the *Front garden2* (with a synonym *polysad (полисад))* and only it bindsto the *Fence 1*. In that case also takes place skipping of value (even obsolete). There takes place also a pass of value (though it is obsolete).

3.4 Skipping of Concepts

1) Friend – loved ones (друг – близкие): The analysis showed that there is a distance of 5 steps between these words in the RuWordNet thesaurus: friend – man – image (human relationships) – quality (attribute, property) – closeness – loved ones (друг – мужчина – образ (человеческиеотношения) – качество (признак, свойство) – близость – близкие). As you can see, the path from an objective noun to a sub-stantiated adjective goes through the concepts of *quality (attribute)* contributing the objectivity transformation into the attributive quality. We did not find the substanti-ated form of the word *loved ones (близкие)* in our sources, but found a synset*close relationships (близкиеотношения)* and the cognate word *близкий* (close), which in Ozhegov dictionary is interpreted through the meaning closely related to the concept of *друг* (friend*)*: 'connected by close personal communication, friendship or love'. This is one of the four meanings recorded in this dictionary. Thus, several ways are possible to solve the problem of bringing the analogues closer. The most optimal one – is the introduction of the concept *loved ones (близкие)* in the meaning of the substantivized adjective 'relatives, friends' [20], which would significantly reduce the distance between the considered analogues.

Another option is to reconsider the semantic relationships between network members in terms of the part-of- speech synonymy of the concepts *friend – loved ones*.

The concept *друг* (friend) in Ozhegov dictionary has 3 meanings: 1. 'A person, who is associated with someone through friendship' (close relations based on mutual trust, affection, common interests); 2. 'The follower, the defender of something or somebody'; 3. Used as an appeal to a loved one, and also (colloq.) as a benevolent appeal in gen-eral [18]. The concept *friend* is presented in the RuWordNet thesaurus in 2 meanings: Friend 1(friend (buddy) hyperonym – *buddy*, further – *acquaintance*; Friend 2 (friend, boyfriend), hyperonym – *male person*.

The concept *близкий* (close) in the thesaurus is presented as a polysemantic adjective: Close 1 (closeness (similarity)); Close 2 (close in distance); Close 3 (close in time); Close 4 (close in relationship).

In order to bring Close 4 (a hyperonym*qualitative*) closer to Friend 1, it is proposed to add a noun *friendship* to the part-of-speech synonymy of the adjective*дружба* (friendship)'. The second variant of bringing closer: add the concept *приятель* (buddy) to the part-of-speech synonymy of synset*close relations (близкиеотношения)*.

So, as a result of our observations, we have determined the following main reasons for the lack of small distance between words in RuWordNet: skipping of the words meaning, referring to different semantic areas, skipping of semantic relations and skipping of concepts. Based on a comparative analysis with the data from the explanatory dictionaries, the NEDS proposed methods for bringing closer of analogues remote in the RuWordNet thesaurus with respect to the semantic relations of inclusion or intersection of meanings. With the help of the component analysis, we identified key words considering their preservation or loss in hypo-hyperonymic relationships. The result of such approach application was the formulation not only of recommendations for RuWordNet for specific words, but also to carefully analyze and take into account the values from the explanatory dictionaries.

At the same time, it was considered that the synset in RuWordNet is a representation of a lexicalized notion (concept), which implements a link with the conceptual categories of consciousness; and also represents the combination of a word with other words, which together form the lexical meaning of the word itself. Application of conceptual analysis enabled to identify significant cognitive features in understanding concepts and go beyond the framework of a systematic approach to a deeper understanding of concepts and the conceptual categories of consciousness.

4 Conclusion

The proposed methodology for computer generation of the shortest paths between words in the thesaurus with subsequent substantial analysis of the path structure is new and can be used for thesaurus verification.

Thus, the semantic relations between analogues in the RuWordNet thesaurus are very diverse, which is reflected in the qualitative and quantitative representation of the adjacency matrix between units of the semantic network.

The basis of distance spacing between words in the RuWordNet thesaurus is the lack of semantic connections between some members of the series, the inconsistency of their routes with the general linguistic representations fixed in the dictionaries. This fact requires a careful study of the semantic relationships of analogues in the RuWordNet thesaurus by bringing them closer together on the basis of a comprehensive analysis assuming not only a conceptual level of semantics, but also its deeper section, including various modifications of the communicative fragment.

Acknowledgments. This research was financially supported by RFBR, grant №. 18-00-01238 and by the Russian Government Program of Competitive Growth of Kazan Federal University.

References

1. Fellbaum, Ch. (ed.): WordNet: An electronic lexical database. MIT Press, Cambridge (1998)
2. Loukachevitch, N., Lashevich, G., Dobrov, B.: Comparing Two Thesaurus Representations for Russian. In: Proceedings of the 9th Global WordNet Conference (GWC 2018), Singapore, pp. 35–44 (2018)
3. Dudareva, Ya.A.: Nominative units with close meaning as components of the associative-verbal network of native speakers of the Russian language. Philological Sciences Ph.D. thesis. Kemerovo (2012) (in Russian)
4. Apresyan, Yu.: Selected Works, Volume I. Lexical Semantics. 2nd edn. School "Languages of Russian Culture", Moscow (1995). (in Russian)
5. Loukachevitch, N.: Thesauruses in Information Retrieval Problems. Publishing house of Moscow State University, Moscow (2011).(in Russian)
6. Apresyan, Y. (ed.): A New Explanatory Dictionary of Synonyms of the Russian Language, 2nd edn. Yazykirusskoykultury, Moscow (2004). (in Russian)
7. Steyvers, M., Tenenbaum, J.B.: The large-scale structure of semantic networks: statistical analyses and a model of semantic growth. Cogn. Sci. **29**(1), 41–78 (2005)
8. Zhu, X., Yang, X., Huang, Y., Guo, Q., Zhang, B.: Measuring similarity and relatedness using multiple semantic relations in WordNet. Knowl. Inf. Syst. **62**(4), 1539–1569 (2019). https://doi.org/10.1007/s10115-019-01387-6
9. Gao, J.B., Zhang, B.W., Chen, X.H.: A WordNet-based semantic similarity measurement combining edge-counting and information content theory. Eng. Appl. Artif. Intell. **39**, 80–88 (2015)
10. Loukachevitch, N.: Corpus-based check-up for thesaurus. In: Proceedings of the 57th Annual Meeting of the Association for Computational Linguistics, pp. 5773–5779. Association for Computational Linguistics, Florence (2019)
11. Bayrasheva, V.: Corpus-based vs thesaurus-based word similarities: expert verification. In: The XX-th International Scientific Conference "Cognitive Modeling in Linguistics" Proceedings, Rostov-on-Don, pp. 56–63 (2019)
12. Bochkarev, V.V, Solovyev, V.D.: Properties of the network of semantic relations in the Russian language based on the RuWordNet data. J. Phys.: Conf. Ser. **1391**(1), Art. № 012052 (2019)
13. Thesaurus of Russian Language RuWordNet. (in Russian). https://ruwordnet.ru/ru. Accessed 10 Mar 2020
14. Cormen, T.H., Leiserson, C.E., Rivest, R.L.: Introduction to Algorithms, 3rd edn. MIT Press, Cambridge (2009)
15. Miller, G.A.: WordNet: a lexical database for English. Commun. ACM **38**(11), 39–41 (1995)
16. Solovyev, V., Gimaletdinova, G., Khalitova, L., Usmanova, L.: Expert assessment of synonymic rows in RuWordNet. In: van der Aalst, W.M.P., et al. (eds.) AIST 2019. CCIS, vol. 1086, pp. 174–183. Springer, Cham (2020). https://doi.org/10.1007/978-3-030-39575-9_18
17. Nikitin, M.V.: Course of linguistic semantics: textbook. Publishing House of the Russian State Pedagogical University named after A.I. Herzen, St. Petersburg (2007). (in Russian)
18. Ozhegov, S.: Russian thesaurus. (in Russian). https://slovarozhegova.ru/. Accessed 10 Mar 2020
19. Shansky, N.M., Ivanov, V.V., Shanskaya T.V.: Brief etymological dictionary of the Russian language. Uchpedgiz, Moscow (1961). (in Russian)
20. Wiktionary. (in Russian). https://ru.wiktionary.org/wiki/gam. Accessed 02 Mar 2020

A Transformation of the RDF Mapping Language into a High-Level Data Analysis Language for Execution in a Distributed Computing Environment

Wenfei Tang[1] and Sergey Stupnikov[2]([envelope]) [iD]

[1] Faculty of Computational Mathematics and Cybernetics,
Lomonosov Moscow State University,
GSP-1, Leninskiye Gory 1-52, 119991 Moscow, Russia
[2] Institute of Informatics Problems, Federal Research Center
"Computer Science and Control" of the Russian Academy of Sciences,
Vavilova st. 44-2, 119333 Moscow, Russia
sstupnikov@ipiran.ru

Abstract. Nowadays scientific data should be FAIR that are Findable, Accessible, Interoperable and Reusable. Reference implementation of FAIR data management principles proposed recently considers RDF as unifying data model and RDF Mapping Language (RML) as the basic language for data integration. This paper is aimed at development of methods and tools for scalable data integration in the frame of this architecture. A mapping from RML into a high-level data analysis language Pig Latin that runs on Hadoop is considered. The mapping is implemented using model transformation technologies. These allows to execute RML programs in the Hadoop distributed computing environment. According to the experimental evaluation RML implementation developed scales w.r.t. data volume and outperforms related implementations.

Keywords: RDF mapping language · Pig Latin · Model transformation · Hadoop

1 Introduction

Scientific data is an important factor in promoting scientific development and knowledge innovations. Data management methods and tools can help people organize, explore and reuse their data effectively. Obstacles on this way appear due to heterogeneity of data represented using various data models and schemas.

In order to effectively manage and reuse scientific data, a famous academic community Force11 in 2014 proposed a set of scientific data management principles called FAIR [13]. According to these principles, scientific data should be *Findable* (identifiable and described by rich metadata to be easy to find), *Accessible* by both humans and machines, *Interoperable* (should be able to be linked or

© Springer Nature Switzerland AG 2021
A. Sychev et al. (Eds.): DAMDID/RCDL 2020, CCIS 1427, pp. 74–91, 2021.
https://doi.org/10.1007/978-3-030-81200-3_6

integrated with other data) and *Reusable* (can be replicated and/or combined in different settings). Soon a novel interoperability architecture [14] was proposed as a reference implementation of FAIR. This architecture enhance the discovery, integration and reuse of data in repositories that lack or have incompatible APIs. It is based on RDF [6] data model and applies the following technologies:

- *LDP(Linked Data Platform)* [10] defining a set of rules for HTTP operations on Web resources, some of which are based on RDF to provide an architecture for reading and writing associated data on the Web.
- *RML(RDF Mapping Language)* [4] aimed to express customized mapping rules from heterogeneous data structures and serializations to the RDF data model.
- *TPF(Triple Pattern Fragments)* [12] interface designed for Linked Data publishing and client-side execution of common types of queries.

LDP provides metadata-level interoperability that allows machines to discover, interpret and link useful metadata. RML and TPF provide data-level interoperability, and they can be combined to provide a common interface for providing FAIR data transformations in a machine-readable manner.

RML is an important foundation for the implementation of FAIR principles. Modularizable RML mappings describes the structure and semantics of RDF graphs. Each RML mapping describes an resource-centric graph, therefore one can make interoperable only on the data he/she is interested in. On the other hand, RML documents are themselves RDF documents, hence RML can be published, discovered, and reused through standard web technologies and protocols. These give it more advantages over XML schema[1] and other similar technologies.

Several implementations of RML have been already developed. RMLMapper[2] runs on a normal Java environment, loads all data in memory, thus could process relatively small datasets. RMLStreamer[3] runs on Apache Flink[4] clusters, thus could process big input files and continuous data streams.

These implementations execute RML rules to generate RDF data. This paper proposes a more flexible approach with RML rules are converted into a high-level data analysis language preserving their semantics. This makes possible to generate RDF data according to RML rules in different distributed computing environments.

Hadoop[5] software framework is chosen as the primary distributed computing platform having in mind that it still holds the largest market share in big data processing. According to the Datanyze[6] Hadoop's current market share exceeds 20%. To make the RML implementation extensible and understandable the high-level language Pig Latin [8] is chosen as an implementation language. Pig Latin

---

[1] https://www.w3.org/XML/Schema.
[2] https://github.com/RMLio/rmlmapper-java.
[3] https://github.com/RMLio/RMLStreamer.
[4] https://flink.apache.org/.
[5] https://hadoop.apache.org/.
[6] https://www.datanyze.com/market-share/big-data-processing--204?page=1.

can be executed on Hadoop, it provides a simple operation and programming interface for massive data-parallel computing. It allows us to focus more on data processing itself than on programming details. Besides, Pig Latin is easy to expand, and the functions of Pig Latin can be easily extended by user-defines functions (UDFs).

The paper proposes a mapping from RML into Pig Latin as well as respective transformation framework applying *Model-driven engineering* approach. At first, metamodels of RML and Pig conforming Ecore meta-metamodel were developed. At second, RML models are extracted from RML textual representation using Xtext[7] framework. At third, the ATLAS Transformation Language (ATL)[8] was applied to implement RML into Pig mapping rules. This allows to transform any RML Ecore model into respective Pig Ecore model. ATL is also applied to convert Pig Ecore models into textual Pig code. Finally, the approach was evaluated and compared with RMLMapper and RMLStreamer. Evaluation of the approach showed much better results than competitors.

The structure of the paper is as follows. Section 2 considers related work. Section 3 gives a brief introduction to RDF, RML and Pig Latin. Section 4 considers the mapping from RML into Pig Latin. Section 5 introduces the transformation framework. Section 6 presents the results of the evaluation. Section 7 summarizes the work.

2   Related Work

Nowadays RDF is quite popular as both scientific and industrial data model. It is a basic technology of Linked Data, so more and more data are published in RDF. RDF was also chosen as unifying data model for the FAIR reference implementation [14]. Obviously to support RDF as a unifying data model the respective methods and tools should be provided.

The mapping approaches to RDF data model include direct mapping, augmented direct mapping and domain semantics-driven mapping [7], etc. Domain semantics-driven mapping are described using mapping languages such as RML and R2RML.

In 2012 the W3C released R2RML [11] mapping language designed to be database-independent, which means one can define how to map a database into RDF without having to consider what the specific database is. Now R2RML has many implementations, such as db2triples[9], Ultrawrap [9] which provide support for a variety of relational databases. However, there are still a lot of data around us that are not stored in a structured form. In order to map this part of data into RDF more comprehensive techniques are required.

In 2014, Anastasia Dimou et al. [4] proposed RDF Mapping Language (RML) based on R2RML. Comparing to R2RML, RML provides a grammar extension

---

[7] https://www.eclipse.org/Xtext/.
[8] https://www.eclipse.org/atl/.
[9] https://github.com/antidot/db2triples.

and a step further in supporting of data source diversity. Using RML, relational databases XML, JSON, SPARQL, etc. can be mapped into RDF data. Two implementations of RML are being supported and updated now that are RMLMapper(see Footnote 2) and RMLStreamer(see Footnote 3). RMLMapper, for instance, is intended to be executed on a single machine, so dealing with large-scale data is problematic.

The focus of this article is extending RML to a distributed computing platform in a flexible way. Pig Latin is chosen as an implementation language allowing to execute RML mappings in Hadoop environment.

3   RDF, RML and Pig

In this section a brief overview of RDF, RML and Pig Latin languages applied in this work are given.

3.1   RDF

RDF (Resource Description Framework)[10] is essentially a data model. It provides a unified standard for describing entities and resources. Briefly speaking, it is a way and means of expressing things. An RDF specification consists of several SPO (Subject, Predicate, Object) triples.

Three important types of values used in RDF should be noted, namely *International Resource Identifiers (IRIs)*, *Blank Nodes* and *Literals*. Literals are strings that may be accompanied with a language tag or a datatype tag like `"english"@en` or `"100"^^xsd:int`. IRIs are enclosed in angle brackets like `<example:iri>`. Blank nodes represent resources for which a URI or literal is not given: `_:node`. The following constraints are applied to SPOs:

1. The Subject of an SPO can be an *IRI* or a *Blank Node.*
2. The Predicate of an SPO is an *IRI.*
3. The Object of an SPO can be an *IRI*, a *Blank Node* or a *Literal.*

In the paper the Turtle syntax[11] is applied to describe RDF specifications. An example below includes five lines, each line is a sequence of terms (subject, predicate, object), separated by whitespaces and ending with a dot. Lines 1 and 2 describe namespace prefix bindings. Line 3 describes an entity of type *"Stop"*. Lines 4 and 5 define the latitude and longitude of the entity.

```
1  @prefix rdf: <http://www.w3.org/1999/02/22-rdf-syntax-ns#>.
2  @prefix ex: <http://example.com/ns#>.
3  <http://airport.example.com/6523> rdf:type "Stop".
4  <http://airport.example.com/6523> ex:latlong
5  <http://loc.example.com/latlong/50.901389,4.484444>.
```

Listing 31. Example of RDF document

---

[10] https://www.w3.org/RDF/.
[11] https://www.w3.org/TR/turtle/.

3.2   RML

RML is a language used to represent custom mappings from heterogeneous data sets in different formats like XML, JSON, CSV into RDF data sets. RML mappings are expressed themselves as RDF graphs and can be written down in Turtle syntax.

An RML mapping refers logical sources to get data from. A logical source can be a base source (any input source or base table) or a view (in case of databases). Data from a source are mapped into RDF using *triples maps*. A triples map is a rule that maps each row of a database, each record of a CSV data source, each element of an XML data source, each object of a JSON data source, etc. into a number of RDF triples. The rule has two main parts: a *subject map* and multiple *predicate-object maps* intended to describe way of generation the respective parts of SPOs. By default, all RDF triples are placed in the default graph of the output dataset. A triples map can contain *graph maps* that place some or all of the triples into named graphs instead.

More formal definitions of main RML constructs are given below according to the [1,3,4]. The same notation is used in the next section to define a general algorithm of the RML mapping into the Pig language:

- An *RML mapping document* can be represented as a pair *mappingDoc=(dr, tm)*, where $dr \in \mathcal{P}(Directive), tm \in \mathcal{P}(TriplesMap)$. This means an RML mapping document consists of directives and triples maps. Here $\mathcal{P}$ means powerset such that $\mathcal{P}(S)$ is a set of all elements of S.
- A *directive* can be represented as a triple *directive=('@prefix' | '@base', prefix?, uriref)*. Here '@prefix' and '@base' are constant strings. The *prefix* is a name, *uriref* is a valid IRIs [5]. Symbol '|' here means alternative, symbol '?' means optionality. An example of directive is
  @prefix rml: <http://semweb.mmlab.be/ns/rml#>.
- A *triples map* can be represented as *triplesMap=(ls, sm, pom)*, where *ls* is a logical source, $sm \in SubjectMap, pom \in \mathcal{P}(PredicateObjectMap)$.
- A *predicate-object map* can be represented as *predicateObjectMap = (pm, om | rom)*, where $pm \in \mathcal{P}(PredicateMap), om \in \mathcal{P}(ObjectMap), rom \in \mathcal{P}(RefObjectMap)$ so that *pm* is a set of predicate maps, *om* is a set of object map, *rom* is a set of reference object maps. More specifically, predicate maps generate the predicates of the RDF triples, object maps and reference object maps generate the objects of the RDF triples.
- A *term map* can be represented as a triple
  termMap = (value, termType, languageTag | datatype), where *value* can be a template or a reference or a constant. A template is a format string that can that can be used to build strings from multiple components and refers elements of data sources (like columns, XML elements etc.) by enclosing them in curly braces("{" and "}"), for example, {/transport/bus/route/stop@id}. A reference is a path that refers to a JSON object, a CSV record etc., like /person/phone/@telNumber. A constant is a constant string that always generates the same RDF term like transit:Stop. *termType* is either IRI, literal or RDF blank node. In brief,

termMap generate subjects, predicates or objects of RDF triples. Subject maps, predicate maps, object maps, graph maps are term maps.

– *Graph maps* are used to generate named graphs or unnamed default graphs. Graph Maps may occur in subject maps or predicate-object map.

An example of a mapping that contains input data (*Venue.json*), an RML mapping (Listing 33) and its output (Listing 34) follows further. The example is extensively used for illustrations in the next Sect. 4.

```
{"venue":[{
    "latitude": "50.901389", "longitude": "4.484444",
    "location": { "continent": "EU", "country": "BE",
        "city": "Brussels" }} ]}
```

Listing 32. Venue.json

```
1  @prefix gn: <http://www.geonames.org/ontology#>.
2  @prefix rr: <http://www.w3.org/ns/r2rml#>.
3  @prefix rml: <http://semweb.mmlab.be/ns/rml#>.
4  @prefix wgs84_pos: <http://www.w3.org/2003/01/geo/wgs84_pos#>.
5
6  <#VenueMapping>
7  rml:logicalSource [
8    rml:source [a dcat:Distribution;
9      dcat:downloadURL <http://www.example.com/Venue.json>]];
10   rml:referenceFormulation ql:JSONPath;
11   rml:iterator "$"];
12 rr:subjectMap [
13   rr:template "http://loc.example.com/city/{$.venue[*].location.
         city}";
14   rr:class schema:City ];
15 rr:predicateObjectMap [
16   rr:predicate wgs84_pos:location;
17   rr:objectMap [
18     rr:parentTriplesMap <#LocationMapping>;
19     rr:joinCondition [
20       rr:child "$.venue[*].location.city";
21       rr:parent "$.venue[*].location.city";]]];
22 rr:predicateObjectMap [
23   rr:predicate geosp:onContinent;
24   rr:objectMap[rml:reference "$.venue[*].location.continent"]];
25
26 <#LocationMapping>
27 rr:subjectMap [
28   rr:template "http://loc.example.com/latlong/{$.venue[*].
         latitude},{$.venue[*].longitude}"];
```

Listing 33. An example of RML mapping

```
1   @prefix gn: <http://www.geonames.org/ontology#>.
2   @prefix geosp: <http://www.telegraphis.net/ontology/geography/
        geography#>.
3   @prefix wgs84_pos: <http://www.w3.org/2003/01/geo/wgs84_pos#>.
4
5   <http://loc.example.com/city/Brussels>,rdf:type,schema:City
6   <http://loc.example.com/city/Brussels>,gn:countryCode,"BE"
7   <http://loc.example.com/city/Brussels>,geosp:onContinent,"EU"
8   <http://loc.example.com/city/Brussels>,wgs84_pos:location,
9   <http://loc.example.com/latlong/50.901389,4.484444>
```

Listing 34. Output of RML Mapping

Listing 33 shows a part of an RML mapping document that produces the desired RDF triples from the input data *Venue.json*. The mapping includes one triples map named *VenueMapping* (Line 6). The triple map has exactly one logical source (*rml:logicalSource*), which refers to the data to be mapped (*Venue.json*) and specifies the data format.

A triples map has only one subject map. A *rr:subjectMap* may have zero or more *rr:class* properties with *IRI* values. The subject map generates subjects according to *rr:template* property. For instance, subject map at lines 12–14 of Listing 33 produces a triple for each path like *$.venue[*].location.city* in *Venue.csv* that is like line 5 of Listing 34, which indicates that *Brussels* is a city. A triple map may include zero or more predicate-object maps. A predicate-object map creates one or more predicate-object pairs from a logical source. Then it is used in conjunction with a subject map to generate RDF triples. Each predicate-object map has one or more predicate maps, object maps or referencing object maps. Both predicate map and object map are term map.

In the example above, the triples map *VenueMapping* has two predicate-object maps. The first one (lines 15–21) defines the location of the city. For example, lines 8–9 Listing 34 indicates the location of *Brussels*. The second one (lines 22–24) defines the continent of a city (line 7 of Listing 34).

3.3   Pig Latin

Pig Latin is an SQL-like data analysis language that can be automatically compiled into MapReduce, Tez or Spark programs to be executed in distributed Hadoop infrastructure. The language is used to analyze large data sets handling structured, unstructured, and semi-structured data.

Pig has a flexible, fully nested data model and allows for complex and non-atomic data types. Atomic values like a string (`'Brussels'`) or a number (`6523`), can be combined into *tuples* like (`'Brussels'`, `6523`) and further in *relations* that are actually *bags* like {(`'Brussels'`, `6253`), (`'Brussels'`, (`'Belgium'`, `'Europe'`))}. A bag can contain duplicates and patterns of contained tuples can be different.

A Pig program consists of statements, and a Pig statement takes a relation as input and produces another relation as output. Here we give a brief comments and short examples to the Pig statements mentioned in the paper. More details on Pig Latin can be found in [8].

- DEFAULT preprocessor statement defines an alias for the constant:
  `%DEFAULT PersonMapping_referenceFormulation'ql:CSV'`
- LOAD deserializes the input data and transform it into the Pig data model. In particular, an UDF can be used to specify a way of deserialization. The output of *LOAD* is a bag containing several tuples. For example,
  ```
  airport = LOAD 'BusStations.csv' USING loadFunc() AS
    (id, city, bus, latitude, longitude);
  ```
- FOREACH applies uniform processing to every tuple of a data set (for instance, generates a target tuple):
  ```
  city = FOREACH airport GENERATE id, city;
  ```
- JOIN stands for a standard equi-joins in Pig. It supports multiple bags and multiple keys at the same time. For example:
  ```
  join_result = JOIN airport by id, city by id;
  ```
- UNION returns the union of two or more bags. For example:
  ```
  union_result = UNION airport, city;
  ```

4 Mapping of RML into Pig Latin

In this section the general principles of mapping RML into Pig Latin are illustrated. An algorithm formalizing the skeleton of mapping is defined in Subsect. 4.1. After that in Subsect. 4.2 mapping of the basic constructs of RML like logical source, subject map, predicate-object map are illustrated by examples.

It should be noted that in order to generate RDF triples in Pig Latin, we have to define a way to represent RDF triples first. In this paper we represent an RDF triple naturally as a tuple *(subject: chararray, predicate: chararray, object: chararray)* like (<http://airport.example.com/Brussels,6523>, `schema:city,"Brussels"`).

4.1 Skeleton Mapping Algorithm

In order to map RML into Pig, a structure of the subjects, predicates, objects, and graphs that are defined in an RML mapping should be exposed. Algorithm 1 RML2Pig is proposed to describe this mapping process.

Here we follow the definition of RML mapping document presented in Subsect. 3.2: m denotes RML mapping document, dr is an element from $\mathcal{P}(Directive)$, tm is an element from $\mathcal{P}(triplesMap)$, ls is a logical source, sm is a subject map, pom is an element from $\mathcal{P}(predicateObjectMap)$. Variable gms denotes a set of graph maps.

Within this algorithm, directives are mapped into Pig using a semantic function *directive2Pig*. Subject maps are mapped into Pig using a semantic function *subjectMap2Pig*.

Algorithm 1. RML2Pig

Input: $m : RML$ *mapping document*
Output: $pc : set of Pig$ *statements*
1: **let** $m = (dr, tm)$
2: $pc \leftarrow$ directive2Pig(dr)
3: **for each** $triplesMap \in tm$ **do**
4:     **let** $triplesMap = (ls, sm, pom)$
5:     $pc \leftarrow pc \cup$ logicalSource2Pig(ls)
6:     $pc \leftarrow pc \cup$ subjectMap2Pig(sm)
7:     $pc \leftarrow pc \cup$ predicateObjectMap2Pig(sm, pom)
8:     $gms \leftarrow gms \cup$ Graph Maps in sm and pom
9: **end for**
10: $pc \leftarrow pc \cup$ graphMap2Pig(gms,pc)

Predicate-object maps are mapped into Pig using a semantic function *predicateObjectMap2Pig*. Graph maps are mapped into Pig using function *graphMap2Pig*. Mentioned semantic functions are illustrated by examples in the next subsection.

4.2   RML Constructs Mapping

In this section the mapping of the basic constructs of RML like logical source, subject map, predicate-object map are illustrated by example presented in Subsect. 3.2.

4.2.1   Logical Source

Semantic of function *logicalSource2Pig* is illustrated in the Table 1.

The Pig code corresponding to *rml:logicalSource* construct includes two statements: *default* and *load*. The preprocessor statement DEFAULT defines a constant JSON string named *VenueMapping_source*, which contains the location and format of *Venue.json*. Then the LOAD statement is used to load data from *Venue.json* into a variable *VenueMapping_data*.

Auxilary UDF R2PLOADSOURCE takes two parameters. The first parameter is *VenueMapping_source* and the second parameter is JSONPath extracted from the RML document. This UDF extracts values according to JSON paths *$.venue[*].location.city*, *$.venue[*].location.continent* and *$.venue[*].location.country* from *Venue.json* document. Considering that a large file can cause out of memory, these values are stored in a temporary file in CSV format. Then it is bound with the variable *VenueMapping_data*. So the variable become to refer a bag of tuples like ((Brussels),(EU),(BE)).

Generally speaking, to form the second parameter of R2PLOADSOURCE, we need to traverse the entire RML document. The JSON paths constituting the parameter value are located in term maps and referencing object maps.

Table 1. The mapping of a logical source

RML	Pig

`<#VenueMapping>`
`rml:logicalSource [`
`  rml:source [`
`    a dcat:Distribution;`
`    dcat:downloadURL`
`      <http://www.example.com/`
`      Venue.json>]];`
`    rml:referenceFormulation`
`    ql:JSONPath;`
`    rml:iterator "$"];`
	`%DEFAULT VenueMapping_source`
	`  '{"iterator":"$",`
	`  "referenceFormulation":`
	`   "ql:JSONPath",`
	`  "distribution":{`
	`   "downloadURL":`
	`    "http://www.example.com/`
	`    Venue.json"}}'`
	
	`VenueMapping_data =`
	` LOAD 'Venue.json' USING`
	`  R2PLOADSOURCE(`
	`   '${VenueMapping_source}',`
	`   '[["$.venue[*].`
	`     location.city"],`
	`    ["$.venue[*].`
	`     location.continent"],`
	`    ["$.venue[*].`
	`     location.country"]]')`
	` AS (f1:tuple(chararray),`
	`  f2:tuple(chararray),`
	`  f3:tuple(chararray));`

So we just need to extract all distinct values of properties *rr:template*, *rml:reference*, *rr:child* and *rr:parent* from the document and combine them in array. For example, in *VenueMapping* the path *$.venue[*].location.city* is got from *rr:template* property of a *rr:subjectMap*.

This example considers data extraction only from JSON file. Note that UDF R2PLOADSOURCE also supports extraction CSV and XML data files from Hadoop Distributed File System, RDF data from SPARQL endpoints, relational data from several DBMS (MySQL, PostgreSQL, and SQLServer).

4.2.2 Subject Map

Semantic function *subjectMap2Pig* is illustrated here. Consider an example of subject map presented in Table 2. It is represented in Pig by the first FORE-ACH statement in the right column of the table. This statement generates a target triple for each tuple of the source collection *VenueMapping_data*. The first element of a triple is a result of a call of an UDF R2PFORMAT[12]. The second and the third elements of the triple are *rdf:type* and *schema:City* constant

---

[12] https://github.com/tangwwwfei/RML2Pig/tree/master/LoadTurtle/src/main/java/r2ps/udf/pig/R2PFORMAT.java.

Table 2. The mapping of a subject map and a predicate-object map

RML	Pig

`<#VenueMapping>`
`rr:subjectMap [`
`  rr:template`
`    "http://loc.example.com/`
`    city/{$.venue[*].`
`    location.city}";`
`  rr:class schema:City ];`
`rr:predicateObjectMap [`
`  rr:predicate`
`    geosp:onContinent;`
`  rr:objectMap [`
`    rml:reference`
`      "$.venue[*].`
`      location.continent"]];`
	`subject =`
	`FOREACH VenueMapping_data`
	`  GENERATE R2PFORMAT('`
	`    http://loc.example.com`
	`    /city/{$.venue[*].`
	`    location.city}',$0),`
	`    'rdf:type',`
	`    'schema:City';`
	`result = UNION result, subject;`
	`objects =`
	`FOREACH VenueMapping_data`
	`  GENERATE R2PFORMAT('`
	`    http://loc.example.com`
	`    /city/{$.venue[*].`
	`    location.city}',$0),`
	`    'geosp:onContinent',`
	`    R2PFORMAT('"%s"',$1);`
	`result = UNION result, objects;`

strings. The UDF R2PFORMAT accepts two parameters: a template chararray and a tuple.

It replaces the content enclosed in curly braces within the template by the value of the second parameter and returns the result. In this case, the second parameter *$0* refers to the first column (*city*) of tuples from *VenueMapping_data* collection. So resulting triples look like (<http://loc.example.com/ city/ Brussels>, `rdf:type`, `schema:City`). Generated RDF triples are stored in the *result* variable. UNION statement is used to combine *result* and *subject* sets.

4.2.3   Predicate-Object Map

Semantic function *predicateObjectMap2Pig* is illustrated here. Consider an example of predicate object map presented in Table 2. It is represented in Pig by the second FOREACH statement in the right column of the table. As in previous subsection, the statement generates target triples in a similar way. Note that *$1* parameter refers to the second column (*continent*) of tuples from *VenueMapping_data* collection. Triples stored in *objects* variable look like (<http://loc. example.com/city/Brussels>, `geosp: onContinent`, `"EU"`).

An example of a predicate object map referencing a triples map (*LocationMapping*) is presented in Table 3. It is represented in Pig by three consecutive FOREACH statements and a JOIN statement.

Table 3. Predicate-object Map with Referencing Object Map.

RML	Pig
rr:predicateObjectMap [rr:predicate wgs84_pos:location; rr:objectMap [rr:parentTriplesMap <#LocationMapping>; rr:joinCondition [rr:child "$.venue[*]. location.city"; rr:parent "$.venue[*]. location.city";]]];	child = FOREACH VenueMapping_data GENERATE R2PFORMAT(' http://loc.example.com/city/ {$.venue[*].location.city}',$0), $0; parent = FOREACH LocationMapping_data GENERATE $0,$2; objects = JOIN child by ($1), parent by ($1); objects = FOREACH objects GENERATE $0, 'wgs84_pos:location', $2; result = UNION result, objects;

Table 4. The mapping of directive.

RML	Pig

@prefix ex:
<http://example.com/ns#>.
	ns = limit result 1;
	ns = FOREACH ns GENERATE
	  '@prefix',
	  'ex:',
	  '<http://example.com/ns#>';
	result = UNION ns, result;

Here tuples from *LocationMapping_data* collection look like (<http://loc.example.com/latlong/50.901389,4.484444>, (50.901389, 4.484444), (Brussels)).

Join condition requires a specified value of the child triple map *VenueMapping* to be equal to a specified value of the parent triple map *LocationMapping*. Tuples in *child* collection look like (<http://loc.example.com/city/Brussels>, Brussels), tuples in *parent* collection look like (<http://loc.example.com/latlong/50.901389,4.484444>, Brussels). Tuples with equal second elements are joined and final tuples in *objects* collection look like (<http://loc.example.com/city/Brussels>, wgs84_pos:location, <http://loc.example.com/latlong/ 50.901389,4.484444>).

4.2.4 Directive.

Semantic function *directive2Pig* is illustrated here. Consider an example of a directive in the Table 4. The directive is just a triple that has to be copied to the target RDF document. So for each directive a separate FOREACH statement is used to put the respective triple into the *result* variable. In this case the triple obviously looks like (@prefix, ex:, $<$http://example.com/ns#$>$).

5   Transformation

The mapping proposed in the previous section is implemented[13] applying Model-driven engineering approach (Fig. 1).

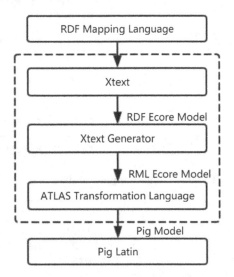

Fig. 1. Modular transformation of RML into Pig

Here the Xtext[14] is a framework that allows us to define a domain-specific language via customized grammar and produce abstract syntax models on the basis of textual representations. A more detailed introduction to the Xtext could be found in [2]. It is assumed that RML documents are initially represented using Turtle[15] RDF textual syntax. So Turtle grammar[16] is used in Xtext to generate *RDF.ecore*[17] that is an RDF abstract syntax metamodel conforming to Ecore meta-metamodel.

So an RML textual document is transformed into an RDF abstract syntax model via Xtext. After that the RDF model is converted into an RML abstract syntax model (*RML.ecore*[18] developed on the basis of RML specification[19]) using

---

[13] RML2Pig Project, https://github.com/tangwwwfei/RML2Pig.

[14] https://www.eclipse.org/Xtext/.

[15] https://www.w3.org/TR/turtle/.

[16] https://www.w3.org/TR/turtle/#sec-grammar.

[17] https://github.com/tangwwwfei/RML2Pig/tree/master/org.xtext.r2ps.rml/model/generated/RML.ecore.

[18] https://github.com/tangwwwfei/RML2Pig/tree/master/RML2Pig/metamodels/RML.ecore.

[19] https://rml.io/specs/rml/.

Xtext Generator[20]. The Xtext Generator is a part of the Xtext allowing us to use a Java-like Xtend[21] language to refine models produced by Xtext.

Abstract syntax of Pig was also formalized as the *Pig.ecore*[22] metamodel conforming Ecore meta-metamodel.

ATL[23] (ATLAS Transformation Language) is used to transform RML models into Pig models. RML into Pig mapping proposed in the previous section was implemented as a set of ATL rules. An example of ATL called rule corresponding to the Algorithm 1 is shown in Listing 51. This called rule *generateMapping* has four *bindings* and each binding invokes a called rule itself. These called rules *GenerateLogicalSource*, *GenerateSubjectMap* and *Namespace2Foreach* correspond to semantic functions *logicalSource2Pig*, *directive2Pig* and *subjectMap2Pig* respectively, and called rule *GenerateAllTermMaps* correspond to semantic functions *predicateObjectMap2Pig* and *graphMap2Pig*.

```
1   rule generateMapping(mapping : RML!Mapping){
2   to program : Pig!Program(
3      statements <- mapping.triplesMaps->
4        collect(t | thisModule.GenerateLogicalSource(t)),
5      statements <- mapping.triplesMaps->
6        collect(t | thisModule.GenerateSubjectMap(t)),
7      statements <- mapping.triplesMaps->
8        collect(t | thisModule.GenerateAllTermMaps(t)),
9      statements <- mapping.directives->
10       collect(e | thisModule.Namespace2Foreach(e))
11  )}
```

Listing 51. A called rule of ATL

6   Evaluation

This section presents a comparative evaluation of the approach with its competitors RMLMapper and RMLStreamer. Testing environment includes AMD Ryzen 7 3700x, 32 GB RAM, and 2 TB HDD with Ubuntu 20.04 operating system and Docker 19.03.

All three implementations were tested using CSV, JSON and XML datasets, which have the same structure as the test data[24] of RMLStreamer. The test datasets were generated by a Python script[25] using a Faker[26] library.

---

[20] https://github.com/tangwwwfei/RML2Pig/tree/master/org.xtext.r2ps.rml/src/ org/xtext/r2ps/rml/generator.
[21] https://www.eclipse.org/xtend/.
[22] https://github.com/tangwwwfei/RML2Pig/tree/master/RML2Pig/metamodels/ Pig.ecore.
[23] https://www.eclipse.org/atl/.
[24] https://figshare.com/articles/Data_for_bounded_data_study/6115049.
[25] https://github.com/tangwwwfei/RML2Pig/tree/master/test-cases/person.py.
[26] https://github.com/joke2k/faker.

Table 5. Evaluation of CSV datasets transformation

CSV	1M	3M	5M	10M	50M
RML2Pig-Tez	41 s	56 s	1 m 38 s	3 m 38 s	16 m 4 s
RML2Scala-Spark	1 m 5 s	1 m 58 s	3 m 4 s	5 m 2 s	23 m 30 s
RMLStreamer	32 s	1 m 31 s	2 m 32 s	5 m 5 s	24 m 51 s
RMLMapper	46 s	2 m 24 s	Error	–	–

Table 6. Evaluation of JSON datasets transformation

JSON	1M	3M	5M	10M	50M
RML2Pig-Tez	48 s	1 m 39 s	2 m 21 s	3 m 42 s	20 m 25 s
RML2Scala-Spark	1 m 8 s	2 m 12 s	3 m 31 s	5 m 55 s	28 m 5 s
RMLStreamer	2 m 29 s	7 m 9 s	11 m 49 s	24 m 38 s	2 h
RMLMapper	59 s	–	–	–	–

Table 7. Evaluation of XML datasets transformation

XML	1M	3M	5M	10M	50M
RML2Pig-Tez	1 m 16 s	2 m 45 s	4 m 1 s	8 m 4 s	42 m 30 s
RML2Scala-Spark	1 m 26 s	3 m 16 s	4 m 58 s	10 m 8 s	49 m 57 s
RMLStreamer	13 m 16 s	Error	Error	Error	Error
RMLMapper	–	–	–	–	–

The data in generated datasets include attributes *PersonId, Name, Phone, Email, Birthdate, Height, Weight, Company*, they correspond to the respective RML mappings[27] (*csv.rml.ttl, json.rml.ttl*, and *xml.rml.ttl*).

The number of records in datasets ranges from 1 million to 50 million. XML data size ranges from 232 MB to 16 GB (Table 7), CSV data size ranges from 130 MB to 4.8 GB (Table 5), and JSON data size ranges from 178 MB to 15 GB (Table 6). The test scripts are available[28].

RMLStreamer runs on Apache Flink clusters, the docker image of Apache Flink cluster[29] is provided by the author of RMLStreamer. It uses one Flink Job Manager and two Flink Task Managers. For the CSV dataset, the *parallelism* parameter of RMLStreamer is set to 4. For the XML and JSON datasets, the *parallelism* parameter have to be equal to 1. The Pig runs on Tez mode on the Hadoop cluster with one NameNode and two DataNodes, the docker image of

---

[27] https://github.com/tangwwwfei/RML2Pig/tree/master/test-cases.

[28] https://github.com/tangwwwfei/RML2Pig/tree/master/TestEnvironment/ TestScripts.

[29] https://github.com/RMLio/RMLStreamer/tree/master/docker.

the Hadoop cluster is available at[30]. The Scala runs on Yarn Cluster mode on Apache Spark with one master node and two slave nodes, the docker images of Spark is available at[31]. RMLMapper also runs on docker[32], it is a standalone environment with OpenJDK 8.

Note that a transformation from RML into Scala (RML2Scala[33]) were also implemented. The mapping is based on the same principles as RML into Pig mapping and share the same UDFs, so the details are omitted in the paper.

As shown in Table 5, the RMLMapper got an error "GC overhead limit exceeded" when the number of records is 5 million. As shown in Table 6, the running time of RMLMapper exceeds 12 h when the number of records is larger than 1 million. So its running time is represented by the symbol "–", which means the test result is unavailable. As shown in Table 7, the RMLStreamer runs for more than 8 h and got an error "Java heap space" when the number of records is 3 million or larger, so its running times are marked as "error". The running time of RMLMapper exceeds 12 h when the number of records is 1 million, so other running times are considered unavailable.

According to the evaluation results, the RMLMapper is the slowest, and has almost no ability to handle large datasets. The RML2Pig on Tez and the RML2Scala on Yarn clusters have better performance than RMLStreamer when processing XML and JSON datasets. And they have slightly better performance than RMLSteamer when processing large CSV dataset. Overall, the running time of our implementations are almost linear, and have better performance than existing implementations. It should be noted that because of the need to provide support for XPath[34], the XML processing largely depends on third-party libraries.

In addition, RML2Pig has passed almost all basic functional tests of RML[35] (331 of 348 cases, i.e. 95%). These test cases are the RML version of the R2RML basic test[36] and are taken from RMLMapper[37] implementation.

7 Conclusions

In this paper we introduced a mapping and transformation framework from the RDF Mapping Language into Pig Latin which enhances RML's ability to

---

[30] https://github.com/tangwwwfei/RML2Pig/tree/master/TestEnvironment/docker-hadoop.
[31] https://github.com/tangwwwfei/RML2Pig/tree/master/TestEnvironment/docker-spark-yarn-cluster.
[32] https://github.com/RMLio/rmlmapper-java.
[33] https://github.com/tangwwwfei/RML2Pig/tree/master/RML2Scala.
[34] https://www.w3.org/TR/xpath-10/.
[35] https://github.com/tangwwwfei/RML2Pig/tree/master/test-cases/resources/passed/test-cases.
[36] https://www.w3.org/TR/rdb2rdf-test-cases/.
[37] https://github.com/RMLio/rmlmapper-java/tree/master/src/test/resources/test-cases.

process large-scale data within distributed computing infrastructures thus supporting reference implementation of FAIR principles. Processing RDF data in a distributed computing environment is the general trend.

It should be noted that some problems have not been solved within this work yet: (1) only a limited variety of data sources are supported[38], (2) some RML features are not supported (function executions, configuration file and metadata generation), (3) only N-Quads[39] output RDF format is supported by now. These problems are considered as a future work. Also other high level languages and computing platforms can be considered to implement RML and compare the performance.

Acknowledgement. The research is financially supported by Russian Foundation for Basic Research, projects 18-07-01434, 18-29-22096. The research was carried out using infrastructure of shared research facilities CKP "Informatics" (http://www.frccsc.ru/ckp) of FRC CSC RAS.

References

1. Anastasia, D., Miel, Vander, S.: RML specification (2014). http://rml.io/spec.html
2. Bettini, L.: Implementing Domain-Specific Languages with Xtext and Xtend. Packt Publishing Ltd. (2016)
3. Dimou, A.: High quality linked data generation from heterogeneous data. Ph.D. thesis, University of Antwerp (2017)
4. Dimou, A., Sande, M.V., Colpaert, P., Verborgh, R., Mannens, E., de Walle, R.V.: RML: a generic language for integrated rdf mappings of heterogeneous data. In: Bizer, C., Heath, T., Auer, S., Berners-Lee, T. (eds.) Proceedings of the Workshop on Linked Data on the Web co-located with the 23rd International World Wide Web Conference (WWW 2014), Seoul, Korea, 8 April 2014. CEUR Workshop Proceedings, vol. 1184. CEUR-WS.org (2014). http://ceur-ws.org/Vol-1184/ldow2014_paper_01.pdf
5. Dürst, M., Suignard, M.: Internationalized resource identifiers (IRIs). Proposed standard RFC 3987, Network Working Group (2005)
6. Klyne, G., Carroll, J.J., McBride, B.: RDF 1.1 concepts and abstract syntax. Recommendation, W3C (2014). https://www.w3.org/TR/rdf11-concepts/
7. Michel, F., Montagnat, J., Zucker, C.F.: A survey of RDB to RDF translation approaches and tools. Research Report hal-00903568 (2014)
8. Olston, C., Reed, B., Srivastava, U., Kumar, R., Tomkins, A.: Pig latin: a not-so-foreign language for data processing. In: Proceedings of the 2008 ACM SIGMOD International Conference on Management of Data. pp. 1099–1110. ACM (2008)
9. Sequeda, J.F., Miranker, D.P.: Ultrawrap mapper: a semi-automatic relational database to RDF (RDB2RDF) mapping tool. In: Villata, S., Pan, J.Z., Dragoni, M. (eds.) Proceedings of the ISWC 2015 Posters & Demonstrations Track

---

[38] Among data retrieval features described in RML documentation JSON, CSV and XML data files from HDFS, RDF data from SPARQL endpoints, relational data from several DBMSs are supported. CSVW (CSV on the Web Vocabulary) and D2RQ Mapping Language are only partially supported. DCAT (Data Catalog Vocabulary) and Hydra core vocabulary for Web API description are not supported.

[39] https://www.w3.org/TR/n-quads/.

co-located with the 14th International Semantic Web Conference (ISWC-2015), Bethlehem, PA, USA, 11 October 2015. CEUR Workshop Proceedings, vol. 1486. CEUR-WS.org (2015). http://ceur-ws.org/Vol-1486/paper_105.pdf

10. Speicher, S., Arwe, J., Malhotra, A.: Linked data platform 1.0. W3C recommendation, W3C, February 2015. http://www.w3.org/TR/2015/REC-ldp-20150226/

11. Sundara, S., Das, S., Cyganiak, R.: R2RML: RDB to RDF mapping language. W3C recommendation, W3C, September 2012. http://www.w3.org/TR/2012/REC-r2rml-20120927/

12. Verborgh, R., et al.: Triple pattern fragments: a low-cost knowledge graph interface for the web. J. Web Semant. **37**, 184–206 (2016)

13. Wilkinson, M.D., et al.: The fair guiding principles for scientific data management and stewardship. Sci. Data **3** (2016). https://doi.org/10.1038/sdata.2016.18

14. Wilkinson, M.D., et al.: Interoperability and fairness through a novel combination of web technologies. PeerJ Comput. Sci. **3**, e110 (2017)

Data Analysis in Medicine

EMG and EEG Pattern Analysis for Monitoring Human Cognitive Activity during Emotional Stimulation

Konstantin Sidorov, Natalya Bodrina$^{(\boxtimes)}$, and Natalya Filatova

Tver State Technical University, Lenina Ave. 25, Tver, Russia

Abstract. The paper describes the experiments on monitoring human cognitive activity with additional emotional stimulation and their results. The purpose of the research is to determine the characteristics of EMG and EEG signals that reflect an emotional state and cognitive activity dynamics. The experiments involved using a multi-channel bioengineering system. The channels for recording EEG signals (19 leads according to the 10–20 system), EMG signals (by the *"corrugator supercilia"* and *"zygomaticus major"* channels according to the Fridlund and Cacioppo methodology) and the protocol information channel were engaged. There is a description of an experimental scenario, which assumed that testees performed homogeneous calculating tasks. According to the experimental results, there were formed 1344 artifact-free EEG and EMG patterns with a duration of 4 s. During emotiogenic stimulation, an EMG signal by the corresponding channel intensifies and a power spectrum shifts to the low-frequency region. An emotional state interpreter based on a neural-like hierarchical structure was used to classify EMG patterns. The classification success was 93%. The authors have determined spectral characteristics and attractors of EEG patterns. The highlighted attractor features were: the averaged vector length for the i-th two-dimensional attractor projection; density of trajectories near its center. The most informative frequency range (theta rhythm) and leads (P3-A1, C3-A1, P4-A2, C4-A2) were selected. These features have revealed a decrease in testees' cognitive activity after 30–40 min of work. After negative emotional stimulation, there was an increase in absolute power in the theta rhythm, an increase in the average vector length for the i-th two-dimensional attractor projection, and a decrease in the trajectory density in four central cells. Tasks success indicators were improving. The revealed EEG signal features allow assessing the current cognitive activity of a person taking into account the influence of emotional stimulation.

Keywords: Cognitive activity · Emotional stimulation · Bioengineering system · Biomedical signal · EEG · EMG · Attractor

1 Introduction

Nowadays, the problem of determining human emotional state is very relevant. It is closely related to the assessment and forecast of operator's performance [1]. One of research areas in this field is studying how emotions affect the level of human cognitive

© Springer Nature Switzerland AG 2021
A. Sychev et al. (Eds.): DAMDID/RCDL 2020, CCIS 1427, pp. 95–109, 2021.
https://doi.org/10.1007/978-3-030-81200-3_7

activity. Much attention is paid to the relationship between human emotional state and the effectiveness of performing cognitive tasks of various types [2–4].

A significant number of experiments in this area have a general scheme, which assumes that a testee has a necessary emotional mood before performing cognitive tasks, which is created by various means [5, 6]. In addition to subjective testee's confirmation of an emotional response, there is a need in quantitative estimates of emotional arousal and cognitive congestion of the brain. For this purpose, the researchers use various biomedical signals (electroencephalogram, electromyogram, electrocardiogram, electrooculogram, photoplethysmogram, etc.) [7–11].

Signals of electrical activity of the brain (EEG) is now considered to be very informative biomedical signal [12–15]. EEG frequency ranges (alpha rhythm and delta rhythm) for tracking fatigue are known [7]. Another paper [8] uses EEG power spectral densities to analyze the effect of image emotional coloring on the effectiveness of solving mnestic tasks. However, there is no clear delineation between signals reflecting emotional state and cognitive activity [8]. For this reason, recording only EEG is not enough. It is necessary to use combined multi-channel systems in studies.

Electromyography (EMG) is widely used to determine the emotional state of a person [16]. The most informative EMG channels (*"corrugators supercilii"* and *"zygomaticus major"*), which make it possible to identify the levels of valence and arousal, are known [9, 17, 18]. The valence of the emotional state ("positive", "negative" or "neutral") is determined with an accuracy of 90%. The accuracy of identification of the six basic senses (happiness, sadness, anger, surprise, fear, disgust) is lower [17–20]. It can be noted that spontaneous emotions are more difficult to identify than simulated emotional responses. To classify emotional states according to EMG, deep learning methods are used: convolutional neural network model [19], L-M BP Neural Network and SWM [20], LSTM and DBN models [17].

While progress has been made in monitoring the emotional state, the level of cognitive activity is still assessed by indirect indicators, which are determined by the results of tasks performing. However, a search for objective methods of assessing cognitive activity for monitoring is being made.

There are known results that describe the activation and interconnection of various structures of the human brain when solving various types of problems [21–23]. In classical psychophysiology, the process of solving problems of the verbal type is associated with the activation of informative indicators of the central and frontal brain zones [24–26]. If a person creates a solution himself without using learned algorithms, then we can observe the activation of the anterior sections of the cerebral cortex associated with regulation [27–29]. There are results proving that the midfrontal theta oscillations involved in cyclic orchestral brain calculations are associated with performing answers during problem solving [23].

The purpose of our research is to highlight the features in order to form an attributive model of cognitive activity monitoring taking into account the effect of emotional stimulation. Using several channels for recording biomedical signals, it is possible to increase the accuracy of the interpreters created on their basis.

For this purpose, it is necessary to form a database with the observation results of EEG and EMG signals of testees who solve homogeneous computing tasks.

From individual fragments of EEG signals $(X(t))$, it is necessary to determine a number of integral characteristics $(F(t))$ that reflect cognitive activity dynamics $(C(A))$:

$$\bigcup_{j=1}^{m} X_j(t) => \bigcup_{i=1}^{r} F_i(t) => C(A) \tag{1}$$

where r is a number of characteristics; m is a number of EEG signal fragments.

$F(t)$ is included in the model under the condition:

$$sign\frac{dF_i(t)}{dt} = sign\frac{dC(A)}{dt} \tag{2}$$

It is also necessary to select the minimum necessary set of electrodes for recording EEG signals.

Using EMG signal, it is necessary to determine the characteristics that confirm an emotional response.

2 The Experiment

A series of experiments involved using a multi-channel bioengineering system "EEG-Speech+" [30, 31]. There were three signal recording channels: EEG channel (Encephalan-131-03 electroencephalograph, Medicom MTD Ltd, Taganrog, Russia), EMG channel (Neuro-MVP-4 myograph, Neurosoft LLC, Ivanovo, Russia) and a protocol information channel.

EEG recording was carried out according to the International 10–20 system [32] by 19 leads with a sampling frequency of 250 Hz.

EMG was recorded on the left side of the face in the "*corrugator supercilia*" and "*zygomaticus major*" channels according to the method of Fridlund and Cacioppo [33]; a ground electrode was placed on the center of the forehead. Recording had a sampling frequency of 1000 Hz.

The participants in the experiments were 7 men and 5 women, who were the employees and students of TSTU at the age of 20–30. The experiments were carried out in a quiet room, during the daytime, the testees sat in a comfortable chair. The duration of each experiment session ranged from 2 to 2.5 h.

During the experiment, a testee performed homogeneous computational tasks (multiplication) in his mind. Each experiment session included 700 tasks. The tasks were presented in groups of 10 units. The complexity of each group was approximately the same. A testee should have said the answer aloud. If an answer was correct, he went on to the next task. The number of attempts per task and the solving time were not limited; missed tasks were not allowed. The protocol had records of all testee's answers (right and wrong) and the time spent on each group of tasks.

The experiment scenario consisted of successive stages (Fig. 1):

I.   Briefing. A testee was instructed on the experimental procedure and on his responsibilities.
II. Training. It did not include connecting of EEG and EMG channels, only information channel recorded signals. This stage involved 20 groups of tasks and allowed a testee to get comfortable.

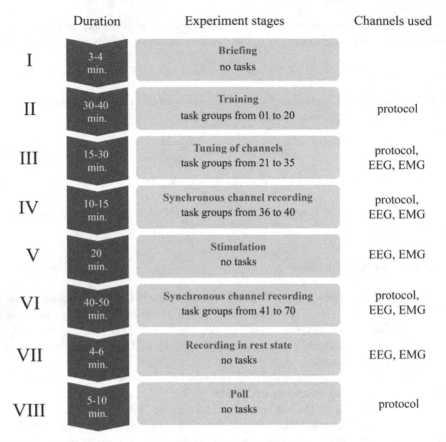

Fig. 1. The experiment scenario

III. The channels were tuned when a testee did groups of tasks from 21 to approximately 35. Researchers placed electrodes on him for long-term recording of biomedical signals, tuned and controlled skin impedance through EEG and EMG channels.

IV. Synchronous channel recording of fell on task groups from 36 to 40 inclusive. By this time, a testee had developed his own mode of operation. By the end of the stage, a testee had been continuously doing tasks for 1.5 h, so the first signs of fatigue began to appear.

V. Stimulation. At this stage, instead of completing tasks, a testee had to focus on the stimulus. In each experiment session, the stimulation included one video clip with a sound lasting 20 min, which could consist of several fragments edited without pauses or interruptions. The theme of the fragments was selected in such a way as to provoke a positive or negative emotional response in a testee. Fragments targeted at obtaining a response of the one sign formed a stimulus. Each testee had two experiment sessions, with positive and negative stimulation.

VI. When a testee did groups of tasks from 41 to 70, recording of channels after stimulation was synchronous.

VII. Recording in rest state. A testee was offered to relax, there were no tasks at this stage. EEG and EMG channels were recorded in a state of quiet wakefulness with closed and open eyes.

VIII. Poll. At the final stage of the experiment, the researchers confirmed that a testee had an emotional response of the required sign at stage V, and found out the response strength. A testee was also asked to evaluate his level of fatigue during the experiment and the complexity of the tasks.

According to the experimental results, the researchers have formed sets of EEG and EMG patterns lasting 4 s each. They selected 56 consecutive time fragments free of artifacts from each experiment session. EEG fragments was taken from stages IV, V, VI, and VII. EMG fragments was taken from stages V and VII.

The obtained EMG records were filtered through a bandpass of 20–500 Hz. Using a notch filter, a network component of 50 Hz was additionally suppressed.

3 Analysis of EMG Patterns

For the analysis, we used a sliding time window without overlapping lasting 4 s. The ratio of the root mean square (RMS_i) for time window i and the average RMS_{sr} fragment with open eyes, when a testee did not do any tasks, was the main sign of emotional changes:

$$RMS_i = \sqrt{\frac{1}{N}\sum_{n=1}^{N}\left|x_n^i\right|^2} \tag{3}$$

where x_n^i is an initial time series of the i-th window of N duration.

Figure 2 shows a comparison of EMG fragments from the *"zygomaticus major"* recording point before, during and after the presentation of an emotional stimulus. When stimulation was positive, the signal from the muscle, which draws the corner of the mouth, was significantly amplified and its spectrum shifted toward low frequencies.

Figure 3 shows a graph of frequency change (median) for the i-th time window, which divides the power spectrum into two parts equal in total intensity:

$$MDF_i = \frac{1}{2}\sum_{k=1}^{M} PSD_k \tag{4}$$

where PSD_k is a power spectral density at k frequency; M is a power spectrum width.

The values of RMS and MDF characteristics are individual and depend on the preferences of a particular person. The effect of the emotional impact, which was manifested in signal amplification and the power spectrum shift, did not last after ending a stimulus and going back to doing cognitive tasks.

In [34], an interpreter of a human emotional state based on a neural-like hierarchical structure (NLHS) was proposed. Emotion interpreter testing was performed through non-crossing training and test sets (Table 1).

Based on training sets analysis a several variants of NLHS was created and classification rules for emotion classes were retrieved. The following results (Table 2) demonstrate the applying of classification rules to signals test sets.

RMS

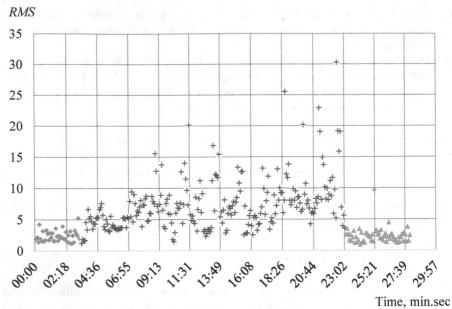

before stimulation   + stimulation   ▲ after stimulation

Fig. 2. Change in the *RMS* in EMG patterns during one experiment session

MDF, Hz

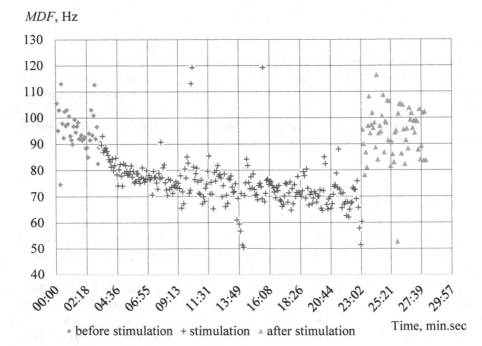

before stimulation   + stimulation   ▲ after stimulation   Time, min.sec

Fig. 3. Change in the *MDF* in EMG patterns during one experiment session

Table 1. The training and test sets structure (EMG)

Set	All	Class 1 – «negative emotion»	Class2 – «positive emotion»
Training set	600	300	300
Test set	744	372	372

Table 2. The results of interpretation for emotions valence (EMG)

Successful classification, %	Training set	Test set
All	100	93
Class 1	100	89
Class 2	100	97

The obtained results (Table 2) demonstrated an acceptable accuracy for classification of EMG patterns. Thus, results of the work prove effectiveness of discrete attributes (3) and (4) generation for emotion's valence differentiation through application of NLHS.

4 Analysis of EEG Patterns

At the first stage of processing EEG signal fragments, we used the methods of spectral analysis. In particular, for all $X(t)$ fragments we determined $S(f)$ power spectra – power spectral densities (PSD_s), using the Fourier transform and Welch's method [35]. We also used the Hamming window (width 512), the frequency range varied from 0 to 125 Hz.

The formalized description of an arbitrary EEG pattern has the following form [36]:

$$S(f)_{EEG} = \langle\{f_1, f_2, \ldots, f_r\}_l\rangle, \tag{5}$$

where $S(f)_{EEG}$ are PSD feature vectors; f is an EEG object number ($f = 1 \div 1344$); l is an EEG lead number ($l = 1 \div 19$); r is PSD feature number ($r = 1 \div 250$), PSD calculation step is 0.5 Hz.

Further, for each calculated feature vector (5), we determined an integral indicator, the so-called absolute power value (AP, $\mu V^2/Hz$) – the area under the corresponding PSD section by selected frequency ranges [37]. This indicator is considered as the total characteristic of the corresponding absolute power fragments (Δf):

$$AP = \bigcup\nolimits_{j=1}^{6} AP_j(\Delta f_j). \tag{6}$$

However, the analysis of the experimental results showed that for features of the form (6), conditions (1) and (2) are fulfilled in no more than 40% of cases.

In our studies, when analyzing EEG patterns and selecting the most informative attribute signs, we used the attractor reconstruction procedure. Based on the reconstruction, we proposed two types of features [4, 38]:

- the averaged vector length ($\bar{R}^i_{max}$) for i-th two-dimensional attractor projection, which is defined as the average value of the 4 maximum vectors in the corresponding plane quadrants (Fig. 4);
- the trajectory density (γ^i_{sum}) in 4 central cells of the grid covering i-th two-dimensional attractor projection (Fig. 4), which is defined as the sum of the points located in the corresponding grid cells.

Experimental studies of the above attractor characteristics have shown that conditions (1) and (2) are fulfilled in more than 60% of cases.

Thus, to assess testee's cognitive activity by one (arbitrary) EEG signal lead, there are 3 feature types:

- Pr1 averaged vector length ($\bar{R}^i_{max}$) for i-th first two-dimensional attractor projection;
- Pr2 trajectory density (γ^i_{sum}) in 4 central cells of the i-th attractor projection;
- Pr3 absolute power AP (Δf) at theta rhythm interval (4–8 Hz).

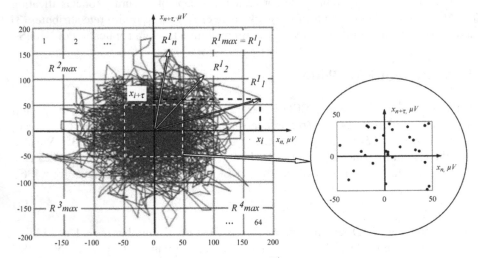

Fig. 4. Determining the length of the averaged vector ($\bar{R}^i_{max} = 175\,\mu V$) and the trajectory density ($\gamma^i_{sum} = 29$) in the four central cells for a two-dimensional projection of the attractor ($x_n - x_{n+\tau}$)

5   The Results EEG Pattern Analysis

The research results made it possible to obtain new data on monitoring levels of human cognitive activity under the conditions of additional emotional stimulation.

During the analysis of absolute power features (6), the most informative EEG leads were localized. They are channels P3-A1, C3-A1, P4-A2 and C4-A2. Particular attention was paid to the selection of useful frequency rhythms illustrating a change in human

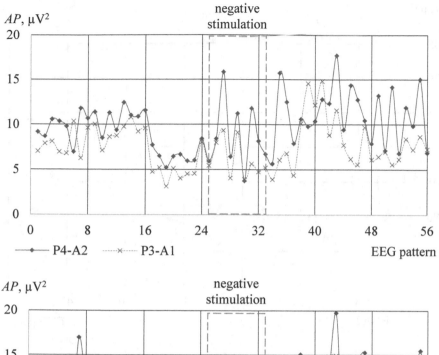

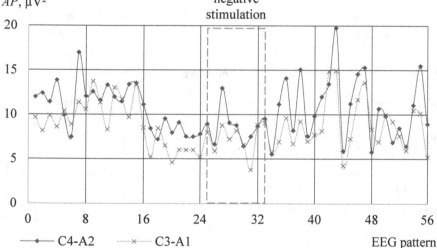

Fig. 5. Signs of absolute power (*AP*, Pr3) in the theta rhythm at a frequency of 4–8 Hz (the shaded area is negative emotional stimulation)

cognitive activity. The experiments showed that the testees had maximum changes in the theta rhythm (Fig. 5).

Monitoring of the EEG signal features (Pr1, Pr2, Pr3) revealed a decrease in cognitive activity at about the 35th–40th minute of the experiment. To increase cognitive activity, we used video stimulation causing a weak negative emotional response.

Theta rhythm after a negative emotional stimulation showed increasing absolute power values (Pr3). The noted pattern maintains this trend for about 30 min.

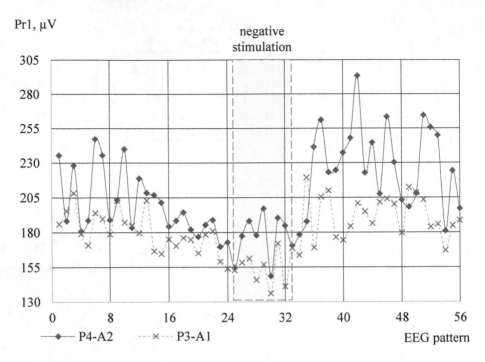

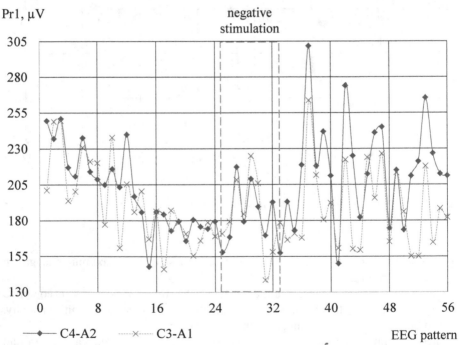

Fig. 6. Signs of length of the averaged vector ($\bar{R}^i_{max}$, Pr1) in the two-dimensional projection of the attractor (the shaded area is negative emotional stimulation)

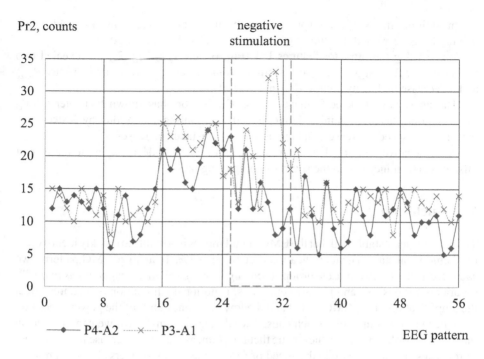

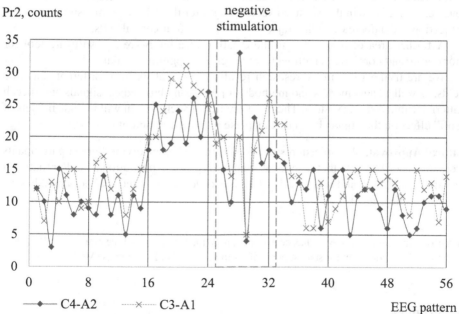

Fig. 7. Signs of a trajectory density (γ_{sum}^{i}, Pr2) in the two-dimensional projection of the attractor (the shaded area is negative e emotional stimulation)

In addition, after stimulation procedures, the number of testees' mistakes made when solving tasks decreased, the time for completing tasks also decreased.

When using the attractor features Pr1 (Fig. 6) and Pr2 (Fig. 7), we revealed two regularities of EEG signals. Emotional stimulation causes an increase in the vector $\bar{R}^i_{max}$ and a decrease in density values γ^i_{sum} in 4 central cells.

The analysis of task performance success indicators has shown that after a long time of doing tasks containing homogeneous computing operations by testees, the performance speed decreases, and the number of mistakes increases.

After negative emotional stimulation, we can observe a decrease in the number of mistakes and an increase in the speed of completing tasks.

6   Conclusion

The research has established that the EMG signal makes it possible to quickly identify testees' emotional state. The identification accuracy is 93%. To analyze EMG patterns, we can effectively use two muscle groups (*"corrugator supercilia"*, *"zygomaticus major"*), separately for positive and negative emotions. Emotional stimulation procedures affect the amplification of the EMG signal and the low-frequency shift of the power spectrum.

After video stimulation, which caused weak negative emotions, we can observe an increase in absolute power values in the theta rhythm, as well as a decrease in the number of mistakes and an increase in the speed of forming testees's answers. The research has revealed an increase in the averaged vector length for the i-th two-dimensional attractor projection and a decrease in the trajectory density in four central cells.

A further area of research is in the creation of a hardware and software tool for monitoring and correcting emotional responses and cognitive activity.

We are trying to combine research results of specialists from different scientific fields, as well as new models and methods of processing biomedical signals and speech patterns within a single system. The functionality of such system will include models of "soft" effect on the human brain that increase its cognitive activity.

Ethical Approval. All procedures performed in studies involving human participants were in accordance with the ethical standards of the institutional and/or national research committee and with the 1964 Helsinki declaration and its later amendments or comparable ethical standards.

Acknowledgements. The work has been done within the framework of the grant of the President of the Russian Federation for state support of young Russian PhD scientists (MK-1398.2020.9).

References

1. Rabinovich, M.I., Muezzinoglu, M.K.: Nonlinear dynamics of the brain: emotion and cognition. Adv. Phys. Sci. **180**(4), 371–387 (2010). https://doi.org/10.3367/UFNr.0180.201004b. 0371. (in Russ., Uspekhi Fizicheskih Nauk)

2. Krutenkova, E.P., Esipenko, E.A., Ryazanova, M.K., Khodanovich, M.Yu.: Emotional pictures impact on cognitive tasks solving. Tomsk State University Journal of Biology **21**(1), 129–145 (2013). (in Russ., Vestnik Tomskogo Gosudarstvennogo Universiteta. Biologiya)
3. Lu, Y., Jaquess, K.J., Hatfield, B.D., Zhou, C., Li, H.: Valence and arousal of emotional stimuli impact cognitive-motor performance in an oddball task. Biol. Psychol. **125**, 105–114 (2017). https://doi.org/10.1016/j.biopsycho.2017.02.010
4. Filatova, N.N., Sidorov, K.V.: Computer models of emotions: construction and methods of research. Tver State Technical University (2017). (in Russ., Kompyuternye Modeli Emotsy: Postroenie i Metody Issledovaniya)
5. Gerjets, P., Walter, C., Rosenstiel, W., Bogdan, M., Zander, T.O.: Cognitive state monitoring and the design of adaptive instruction in digital environments: lessons learned from cognitive workload assessment using a passive brain-computer interface approach. Front. Neurosci. Hypothesis Theory Article. **8**(385), 1–21 (2014). https://doi.org/10.3389/fnins.2014.00385
6. Sidorov, K.V., Filatova, N.N., Shemaev, P.D., Bodrina, N.I.: Application of fuzzy statements for interpretation of the emotional influence on human cognitive activity. Fuzzy Syst. Soft Comput. **13**(2), 147–165 (2018). https://doi.org/10.26456/fssc47. (in Russ., Nechetkie Sistemy i Myagkie Vychisleniya)
7. Pomer-Escher, A., Tello, R., Castillo, J., Bastos-Filho, T.: Analysis of mental fatigue in motor imagery and emotional stimulation based on EEG. In: Proceedings of the XXIV Brazilian Congress of Biomedical Engineering "CBEB 2014", Uberlandia, Brazil, pp. 1709–1712 (2014). https://www.researchgate.net/publication/265207783.
8. Grissmann, S., Faller, J., Scharinger, C., Spuler, M., Gerjets, P.: Electroencephalography based analysis of working memory load and affective valence in an n-back task with emotional stimuli. Front. Hum. Neurosci. **11**(616), 1–12 (2017). https://doi.org/10.3389/fnhum.2017.00616
9. Chołoniewski, J., Chmiel, A., Sienkiewicz, J., Hołyst, J., Kuster, D., Kappas, A.: Temporal Taylor's scaling of facial electromyography and electrodermal activity in the course of emotional stimulation. Chaos Solitons Fractals **90**, 91–100 (2016). https://doi.org/10.1016/j.chaos.2016.04.023
10. Mavratzakis, A., Herbert, C., Walla, P.: Emotional facial expressions evoke faster orienting responses, but weaker emotional responses at neural and behavioural levels compared to scenes: a simultaneous EEG and facial EMG study. NeuroImage **124**, 931–946 (2016). https://doi.org/10.1016/j.neuroimage.2015.09.065
11. Shu, L., et al.: A review of emotion recognition using physiological signals. Sensors **18**(7), 2074 (2018). https://doi.org/10.3390/s18072074
12. Panischeva, S.N., Panischev, O., Demin, S.A., Latypov, R.R.: Collective effects in human EEGs at cognitive activity. J. Phys.: Conf. Ser. **1038**, 012025 (2018). https://doi.org/10.1088/1742-6596/1038/1/012025
13. Montgomery, R.W., Montgomery, L.D.: EEG monitoring of cognitive performance. Phys. Med. Rehabil. Res. **3**(4), 1–5 (2018). https://doi.org/10.15761/PMRR.1000178
14. Magosso, E., De Crescenzio, F., Ricci, G., Piastra, S., Ursino, M.: EEG alpha power is modulated by attentional changes during cognitive tasks and virtual reality immersion. Comput. Intell. Neurosci. 7051079 (2019). https://doi.org/10.1155/2019/7051079
15. Friedman, N., Fekete, T., Gal, K., Shriki, O.: EEG-Based prediction of cognitive load in intelligence tests. Front. Hum. Neurosci. **13**(191), 1–9 (2019). https://doi.org/10.3389/fnhum.2019.00191
16. Perdiz, J., Pires, G., Nunes, U.J.: Emotional state detection based on EMG and EOG biosignals: a short survey. In: Proceedings of 5th Portuguese Meeting on Bioengineering (ENBENG), pp. 1–4. IEEE. Coimbra (2017).https://doi.org/10.1109/ENBENG.2017.7889451

17. Abtahi, F., Ro, T., Li, W., Zhu, Z.: Emotion analysis using audio/video, EMG and EEG: a dataset and comparison study. In: Proceedings of Winter Conference on Applications of Computer Vision (WACV), pp. 10–19. IEEE. Lake Tahoe (2018). https://doi.org/10.1109/WACV.2018.00008

18. Jerritta, S., Murugappan, M., Wan, K., Sazali, Y.: Emotion recognition from facial EMG signals using higher order statistics and principal component analysis. J. Chin. Inst. Eng. **37**(3) (2013). https://doi.org/10.1080/02533839.2013.799946

19. Lee, M., Cho, Y., Lee, Y., Pae, D., Lim, M., Kang, T.: PPG and EMG based emotion recognition using convolutional neural network. In: Proceedings of the 16th International Conference on Informatics in Control, Automation and Robotics (ICINCO), Prague, vol. 1, pp. 595–600 (2019). https://doi.org/10.5220/0007797005950600

20. Yang, S., Yang, G.: Emotion recognition of EMG based on improved L-M BP neural network and SVM. J. Softw. **6**(8), 1529–1536 (2011)

21. Hsu, Y.-F., Xu, W., Parviainen, T., Hämäläinen, J.A.: Context-dependent minimization of prediction errors involves temporal-frontal activation. NeuroImage **207**, 116355 (2020). https://doi.org/10.1016/j.neuroimage.2019.116355

22. Ouyang, G., Hildebrandt, A., Schmitz, F., Herrmann, C.S.: Decomposing alpha and 1/f brain activities reveals their differential associations with cognitive processing speed. NeuroImage **205**, 116304 (2020). https://doi.org/10.1016/j.neuroimage.2019.116304

23. Duprez, J., Gulbinaite, R., Cohen, M.X.: Midfrontal theta phase coordinates behaviorally relevant brain computations during cognitive control. NeuroImage **207**, 116340 (2020). https://doi.org/10.1016/j.neuroimage.2019.116340

24. Gray, J.R., Braver, T.S., Raichle, M.E.: Integration of emotion and cognition in the lateral prefrontal cortex. Proc. Natl. Acad. Sci. U.S.A. **99**(6), 4115–4120 (2002). https://doi.org/10.1073/pnas.062381899

25. Thayer, J.F., Hansen, A.L., Saus-Rose, E., Johnsen, B.H.: Heart rate variability, prefrontal neural function, and cognitive performance: the neurovisceral integration perspective on self-regulation, adaptation, and health. Ann. Behav. Med. **37**(2), 141–153 (2009). https://doi.org/10.1007/s12160-009-9101-z

26. Kropotov, J.: Quantitative EEG, Event-Related Potentials and Neurotherapy, 1st edn. Academic Press, London (2009)

27. Simonov, P.V. The Emotional Brain. Nauka Publ., Moscow (1981). (in Russ., Emocionalnij mozg)

28. Baldwin, C.L., Penaranda, B.N.: Adaptive training using an artificial neural network and EEG metrics for within- and cross-task workload classification. NeuroImage **59**(1), 48–56 (2012). https://doi.org/10.1016/j.neuroimage.2011.07.047

29. Smirnitskaya, A.V., Vladimirov, I.Yu.: Differences in the activity of the executive functions in algorithmic and insight problem solving: ERP study. Steps **3**(1), 98–108 (2017). (in Russ., Shagi)

30. Filatova, N.N., Bodrina, N.I., Sidorov, K.V., Shemaev, P.D.: Organization of information support for a bioengineering system of emotional response research. In: Proceedings of the XX International Conference "Data Analytics and Management in Data Intensive Domains" DAMDID/RCDL. CEUR Workshop Proceedings, pp. 90–97. CEUR. Moscow, Russia (2018). http://ceur-ws.org/Vol-2277/paper18.pdf

31. Filatova, N.N., Sidorov, K.V., Shemaev, P.D., Rebrun, I.A.: Emotion and cognitive activity monitoring system. In: Proceedings of the 3rd Russian-Pacific Conference on Computer Technology and Applications "RPC 2018", pp. 1–4. IEEE. Vladivostok, Russia (2018). https://doi.org/10.1109/RPC.2018.8482220

32. Jasper, H.H.: The ten-twenty electrode system of the international federation. Electroencephalogr. Clin. Neurophysiol. **10**, 371–375 (1958)

33. Fridlund, A.J., Cacioppo, J.T.: Guidelines for human electromyographic research. Psychophysiology **23**(5), 567–589 (1986). https://doi.org/10.1111/j.1469-8986.1986.tb00676.x
34. Sidorov, K., Filatova, N., Shemaev, P.: An interpreter of a human emotional state based on a neural-like hierarchical structure. In: Abraham, A., Kovalev, S., Tarassov, V., Snasel, V., Sukhanov, A. (eds.) IITI'18 2018. AISC, vol. 874, pp. 483–492. Springer, Cham (2019). https://doi.org/10.1007/978-3-030-01818-4_48
35. Rangayyan, R.M.: Biomedical Signal Analysis. 2nd edn. Wiley-IEEE Press, New York (2015). https://doi.org/10.1002/9781119068129
36. Sidorov, K.V., Filatova, N.N., Bodrina, N.I., Shemaev, P.D.: Analysis of biomedical signals as a way to assess cognitive activity during emotional stimulation. Proc. Southwest State Univ. Ser.: Control Comput. Eng. Inf. Sci. Med. Instr. Eng. **9**(1), 74–85 (2019). (in Russ., Izvestiya YUgo-Zapadnogo Gosudarstvennogo Universiteta. Seriya: Upravleniye, Vychislitelnaya tekhnika, Informatika)
37. Klimesch, W.: EEG alpha and theta oscillations reflect cognitive and memory performance: a review and analysis. Brain Res. Rev. **29**(2–3), 169–195 (1999). https://doi.org/10.1016/S0165-0173(98)00056-3
38. Filatova, N.N., Sidorov, K.V., Shemaev, P.D., Iliasov, L.V.: Monitoring attractor characteristics as a method of objective estimation of testee's emotional state. J. Eng. Appl. Sci. **12**, 9164–9175 (2017)

Finding the TMS-Targeted Group of Fibers Reconstructed from Diffusion MRI Data

Sofya Kulikova$^{(\boxtimes)}$ ⓘ and Aleksey Buzmakov ⓘ

National Research University Higher School of Economics, Perm, Russia
SPKulikova@hse.ru

Abstract. Transcranial magnetic stimulation is considered as a promising diagnostic and therapeutic approach, despite the fact that its mechanisms remain poorly understood. Theoretical models suggest that TMS-induced effects, within brain tissues, are rather local and strongly depend on the orientation of the stimulated nervous fibers. Using diffusion MRI, it is possible to estimate local orientation of the white matter fibers and to compute effects, that TMS impose at each point of them. The computed effects may be correlated with the experimentally observed TMS effects. However, since TMS effects are rather local, such relationships are likely to be observed only for a small subset of the reconstructed fibers. In this work, we present an approach for finding such a TMS-targeted subset of fibers, within a cortico-spinal tract, following stimulation of the motor cortex. Finding TMS-targeted groups of fibers is an important task for both (1) better understanding of the neuronal mechanisms, underlying the observed TMS effects and (2) development of future optimization strategies for TMS-based therapeutic approaches.

Keywords: Transcranial magnetic stimulation · Diffusion MRI · Target fibers

1 Introduction

Transcranial magnetic stimulation (TMS) presents a potentially powerful diagnostic and therapeutic approach for various neurological and psychiatric states [7]. TMS is based on changing neuronal activity in the brain cortex, following application of electromagnetic pulses, generated by a stimulation coil. At the whole organism level, TMS-related effects may be observed in various forms, depending on the site of stimulation, as well as on other stimulation parameters, such as stimulation intensity, pulse shape, delays between subsequent pulses, etc. For example, stimulation of the motor cortex typically manifests in muscle twitches that could be recorded using electromyography [8]. Repetitive TMS of language-related areas interfere with speech production and is commonly used for pre-surgical planning [3]. Furthermore, stimulation of higher-order associative areas, such as dorso-lateral prefrontal cortex (DLPFC), can even change the processes of decision-making [9].

© Springer Nature Switzerland AG 2021
A. Sychev et al. (Eds.): DAMDID/RCDL 2020, CCIS 1427, pp. 110–121, 2021.
https://doi.org/10.1007/978-3-030-81200-3_8

However, the exact neuronal mechanisms, that account for the observed TMS-induced effects, remain largely unknown. Furthermore, TMS effects are prone to high inter-subject variability, which limits TMS usage in clinical practice. Existing theoretical models suggest that TMS effects, within brain tissues, are rather local and depend on the local geometry of the stimulated nervous fibers [5, 19]. Thus, the experimentally-observed inter-subject variability of the TMS-induced effects may arise from individual differences in the anatomical organization of the white matter. Although, to the best of our knowledge, there were no experimental studies directly testing these theoretical models, several previous works indicated that even changing the orientation of the stimulating coil at the same stimulation site and thus, changing the orientation of induced electric field relative to the stimulated nervous fibers, may significantly affect the observed TMS-induced effects [15].

Non-invasive investigation of the white matter fibers became possible with the introduction of the diffusion-weighted magnetic resonance imaging (MRI) [11]. Diffusion MRI measures diffusion (brownian notion) of the water molecules within brain tissues. When the medium is isotropic (for example, within the brain ventricles or in the grey matter), water molecules displace in any direction with equal probability. However, within the regions, containing highly arranged structures, such as nervous fibers, molecular displacement is restricted to the direction of the local fiber orientation. Thus, by measuring diffusion in multiple directions, it is possible to reveal the principal diffusion direction(s), reflecting the orientations of the fibers, passing through each point. Then, tractography algorithms may be applied to perform 3D-reconstructions of the white matter fibers. The reconstructed fibers are represented by streamlines that reflect typical trajectories of the fibers.

Combining experimental TMS data with theoretically computed TMS effects, that take into account local geometrical characteristics of the white matter fibers, may be used to challenge the theoretical models and to get new insights on the neuronal mechanisms, underlying TMS effects. Since TMS effects are likely to be local, the computed effects should be related to the experimental data only for a small subset of the reconstructed streamlines. The question on how to find this small subset of the streamlines is related to subgroup discovery task [1]. Unfortunately, there is no subgroup discovery method searching for the best subgroup w.r.t. correlation. Thus, in this paper we define a special brute force approach for finding TMS-targeted group of the reconstructed streamlines. Accordingly, the contribution of the paper is two-folds: (1) the new approach for finding the TMS-targeted group and (2) the findings suggesting that the muscle related fibers are localized in a small area.

Of importance, finding the TMS-targeted groups of fibers should improve our understanding of the neuronal mechanisms guiding the observed TMS-related effects and make important contribution to the optimization of TMS-approaches including their clinical applications.

The rest of the paper is organised as following. First, in Sect. 2 we introduce the used methods. Then, Sect. 3 presents the obtained results. Finally, before concluding the paper, the results are discussed in Sect. 4.

2   Methods and Materials

2.1   The Study Pipeline

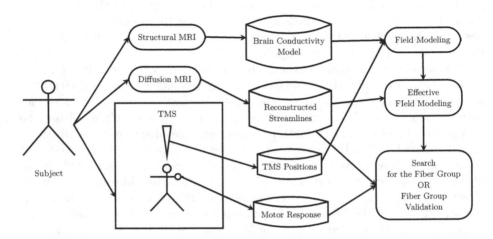

Fig. 1. The overall study pipeline

In this work, we used TMS and MRI data of one healthy subject from the study of Novikov et al. [13]. The overall pipeline of the study is shown in Fig. 1. The subject has undergone structural MRI session in order to get a T1-weighted image of the brain. This image was used to create the brain conductivity model that was further used in the electric field modeling. The subject has also undergone the diffusion MRI session that was necessary for subsequent reconstructing of the streamlines. Finally, the subject has undergone the motor TMS session and the motor responses from a right hand were registered with the corresponding the position of the stimulation coil relative to the subject's head.

Further, the TMS coil positions and the brain conductivity model were used to model the electric field inside the brain. However, the effective field along the nerve fibers is likely to be different than the surrounding induced field. Thus, the effective field was computed based on streamline geometry and the surrounding electric field using in-house developed software StimVis [10].

Every streamline was associate with the maximal value of the effective field along the streamline. This maximal effective field was used to decide whether the streamline is activated or not. Finally, depending on the experimental day this data was used either for finding the targeted group of fibers or for validating the

found result. The search for the targeted group of fibers was done by iterating
through the positions of the streamlines and through the activation thresholds,
maximizing the correlation between the number of activated streamlines and the
corresponding amplitude of the motor response.

2.2 Description of the Experimental Data

MRI data was acquired on a 1.5T Siemens scanner (Siemens Healthcare,
Erlangen, Germany) and included a structural T1-weighted image and a
sequence of diffusion-weighted images. T1-weighted image was obtained with
a 1mm isotropic spatial resolution using a 3D fast gradient inversion recovery
sequence (MPRage). Diffusion-weighted images were acquired using a DW-SE-
EPI sequence with the following parameters: 64 diffusion gradient orientations
($b = 1500$ s/mm^2), 4 b0-images, $TE = 72$ ms, $TR = 14$ s, voxel dimensions equal
to $1 \times 1 \times 2$ mm, parallel imaging GRAPPA factor 2, partial Fourier sampling
factor 6/8.

TMS sessions were performed using a Nexstim eXimia stimulator with nTMS-
compatible electromyography (EMG) device and a figure-of-eight coil (Focal
Bipulse, Nexstim Plc, Helsinki, Finland). Two TMS motor mapping sessions
were performed on two different days (Day 1 and Day 2) separated by 7 days.
Coordinates of the stimulation sites were chosen according to a virtual MRI-
based grid and were stimulated in a pseudo-random order. Stimulation sites on
Day 2 followed exactly the same order as on the Day 1. Navigation error for
each stimulation site was kept below 2 mm. In this study we considered the first
55 stimulation sites from Day 1 and the corresponding 55 stimulation sites from
Day 2. Motor Evoked Potentials (MEP) were measured for right hand muscles
and MEP amplitudes were calculated automatically using the eXimia software.

2.3 Data Pre-processing

To compute distributions of the TMS-induced electric fields within brain tissues,
using SimNIBS software [18], it is necessary to create a volume conductor model
(VCM) of the head from a structural T1-weighted image. This step was done
using an mri2mesh algorithm [23]. Shortly, this algorithm relies on segmenting
structural MRI images into 5 tissue classes (white matter, grey matter, skin, scull
and cerebral spinal fluid), subsequent volume meshing and assigning appropriate
conductivity values [20,22] to each mesh node. Once the VCM file is computed,
it could imported to SimNIBS [18] to simulate the TMS-induced electric fields
for each of the experimental stimulation sites. Simulation results are stored in
the form of a head mesh with electric field vector, assigned to each node of the
mesh.

To perform tractography, the diffusion data was pre-proccessed with a standard pipeline in Diffusion Toolkit[1]. A whole-brain tractography and FA maps

---

[1] http://www.trackvis.org.

were computed according to a Diffusion Tensor Model [11]. The resulting stream-lines were visualized in TrackVis (see footnote 1) software and the cortico-spinal tract (CST), the principal descending pathway that conveys motor information, was extracted by manually placed regions of interest (ROI). Diffusion MRI-derived data were co-registered with TMS-related data using affine transformation that maximizes the mutual information between T1-weighted image and FA maps. Co-registration is performed in 4 steps and the resulting transformation matrix at each step is used as a starting matrix for the optimization at the next step: 1) calculation of the center of mass transform; 2) calculation of a 3D translation transform; 3) calculation of a 3D rigid transform; 4) calculation of an affine transform. Both optimization and subsequent data transformation relies on the algorithms implemented in DIPY (Diffusion Imaging in Python) [4].

2.4   Calculation of the TMS-Induced Effects

TMS-induced effects were computed for each point on each CST streamline using an in-house developed software StimVis[2] [10] which implements previously suggested theoretical models of axonal excitation by the electromagnetic fields [5,19]. According to these models, TMS-induced effects depend on the effective electrical field, i.e. projection of the induced electrical field E_l within the brain tissues to the local orientation of the stimulated axons. Three distinct excitation mechanisms are thought to contribute to the total TMS-induced effect. The first mechanism follows from the cable equation [16] and describes an electromagnetic induction on the membrane potential V relative to its resting value. The magnitude of the corresponding effects equals to $-\lambda^2 \frac{\partial E_l}{\partial l}$ (where λ is the length constant of the neural membrane equal to 2 mm). The second mechanism is proportional to the magnitude of the effective field $-\lambda E_l$ and mainly reveals itself at sharp bends and terminations of the fibers. Finally, the third mechanism may happens at the interface between different tissue types due to a steep change in conductivity values [12]. However, the later mechanism is rather rare and its effects are at least an order of magnitude smaller that from two other mechanisms [5]. Thus, in the present work we considered only effects related to the first two excitation mechanisms.

To calculate the TMS-related effects as described above, one should first get an electric field vector E at each point on each CST streamline. To do so for a given point, a tetrahedra containing that point is found withing the head mesh, generated by SimNIBS. Each node of the tetrahedra is associated with previously computed electric field vector. Then, the electric field vector at the point of interest is obtained by linear interpolation from the nodes of that tetrahedra. Once electric field is known, it is projected to the local direction of the streamline l to get the effective field E_l. The directional derivative of the effective field is calculated as a dot product between the gradient of the effective field E_l and the unit vector, tangent to the streamline: $\frac{\partial E_l}{\partial l} = \nabla E_l \cdot l$. Finally,

---

[2] https://github.com/KulikovaSofya/StimVis_TMS.

the total TMS-induced effect is obtained as a sum of the effects from the two excitation mechanisms: $-(\lambda E_l + \lambda^2 \frac{\partial E_l}{\partial l})$.

2.5   Finding TMS-Targeted Groups of Fibers

The TMS stimulation is performed over a relatively large area. However, the motor response is registered only from a single muscle. Accordingly, in order to connect the theoretical effect of the effective field in a streamline and the registered motor response, it is necessary to determine which streamlines are related to this motor response. In what follows it is assumed that the streamlines responsible for similar behaviour are placed together in the space. Accordingly, the task is to find closely placed streamlines, such that their activation correlates with the registered motor response.

Close CST Streamlines. Most of the CST streamlines span from the center to the top of the brain (see Fig. 3), accordingly the streamlines can be considered to be close to each other if their projection to horizontal plain is close to each other. We associate every streamline with the average coordinate in the horizontal plain for the points of the middle part of the streamline (shown in red in Fig. 3), since this part is quite straight, the average coordinate of the streamline is a good proxi for the streamline spatial position.

Streamline Activation. The next question is how to determine that a streamline is activated. One of the activation hypothesis is that, if the maximal electric field induced on the streamline is higher than a certain threshold, then the streamline is activated. However, the threshold is not known. Thus, we need not only to find a good set of streamlines but also the activation threshold for streamlines.

Quality Function for the Group of Streamlines. Accordingly, we need to find a group of closely placed streamlines, the corresponding threshold activation and then to relate the activation of the group to the registered motor response. The activation level of the group is measured as the number of the activated streamlines, i.e., the streamline with the induced electrical field higher than a threshold, within the group. For every group of streamlines and every threshold, every physical position of the TMS coil induce different electric field over the streamlines and, consequently, differently activate the group of streamlines. Thus, in order to estimate how well the selected group of streamlines is related to the registered motor response, we correlated the number of the activated streamlines (w.r.t. the threshold) in the group with the amplitude of the motor response. The higher the correlation is the more related streamlines are within the group. We used Spearman correlation since there is no reason to assume a linear relation between the number of the activated streamlines and the motor response.

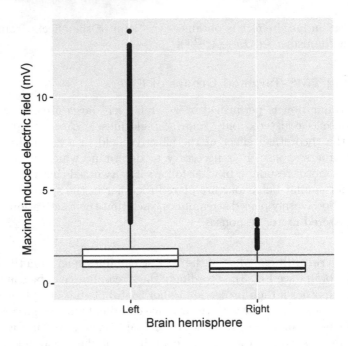

Fig. 2. Distribution of the induced electric fields for different streamlines in different brain hemispheres and different TMS positions

Search for the Best Group of Streamlines. Every streamline was associated with a coordinate in horizontal plain and for every position of TMS, it was associated with the maximal induced electric field. Then, the search for the best group of streamlines is formalized as following: find the best rectangle in the horizontal plain and the best threshold, such that the number of activated streamlines having the coordinate within the rectangle is correlated with the registered motor response.

This task can be solved by a brute force approach. However, in order to make it computable, one needs to add some simplification. First, we put the coordinate system in the horizontal plane such, that x-axis is placed from the left to the right of the head and the y-axis is placed in the caudo-rostral direction. Then, we are interested only in the rectangles that are parallel to the axes. Finally, for every rectangle, we allow only 10 possible positions for every edge, making it 10^4 possible rectangles in total. These 10 positions divide the streamlines to 10 equally-sized groups w.r.t. both axes.

For finding the best threshold, we compared the distribution of the induced electric field in the left and in right CST streamlines (see Fig. 2). It should be noticed, that only one of these two bundles (the left CST) was stimulated. Accordingly, the first threshold was selected to be higher than most of the values of the induced electric field in the right bundle of the CST streamlines. Then, all higher values for the left bundle of streamlines were divided into 5 equally-sized groups, thus, we examined 5 possible activation thresholds.

Group Validation. Such a brute force approach introduce multiple hypothesis testing, thus the found result should be validated on an independent sample of the data. For such validation, we used data from another experimental day, where similar TMS positions were used for the same subject. The previously found group of streamlines and the corresponding activation threshold are considered as the only hypothesis. Accordingly, for the new TMS possitions the electric field was modeled. Then, for only the found group of streamlines the number of the activated streamlines (w.r.t. the new electric field and the previously found activation threshold) was computed. This number was correlated with the newly registered motor response. Since in the second experiment there is only one hypothesis, no multiplicity adjustment was needed.

3   Results

The reconstructed cortico-spinal tract contained 1358 streamlines, almost equally splitted between the two hemispheres (Fig. 3). For every point of the streamlines, the induced field was computed for every TMS position. Given a TMS position, every streamline was associated with the maximal induced electric field along the streamline. In Fig. 2 the distribution of the maximal induced electric field is shown in left and right hemispheres of the brain. It can be seen that the target hemisphere is associated with the higher induced electric field. The level 1.5 mV (shown by horizontal line in Fig. 2) was further used as the minimal activation threshold.

First, the brute force search was applied to the TMS data of the first experimental day. The best found group of streamlines corresponds to the minimal activation threshold equal to 1.5 mV. It is placed at the most left positions w.r.t. the x-axis and in the middle of the y-axis. Totally, 12 streamlines were found (Fig. 4). The Spearman correlation between the number of activated streamlines within that group and the observed motor response is equal to 0.51 (p-value is $6 \cdot 10^{-5}$).

This result was validated on the second day for the same set of streamlines. The corresponding Spearman correlation was 0.37 (p-value is $0.5 \cdot 10^{-2}$).

4   Discussion

To the best of our knowledge, this study presents the first attempt to directly test the relationships between the existing theoretical models of TMS mechanisms and the observed experimental TMS effects. We exploited motor TMS for several reasons. First of all, its effects could be easily and quickly measured by amplitudes of the motor evoked potentials in the relevant muscles. Second, existing somotopy of the motor cortex [6] facilitates navigation during TMS

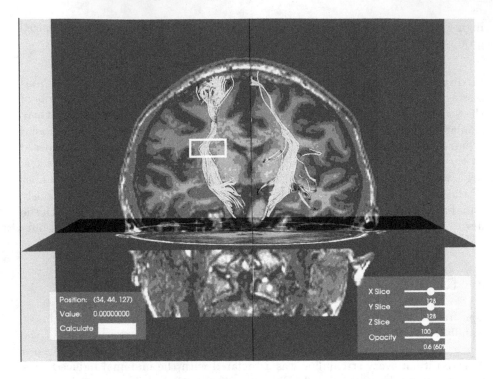

Fig. 3. CST streamlines reconstructed from DTI data.

sessions. Third, motor TMS effects are related to the excitation of the descending motor neurons whose axons form the cortico-spinal tract. This tract has a relatively simple shape that can be successfully reconstructed even with very simple diffusion models, such as a diffusion tensor model. And finally, preserved somatotopy withing the brain motor system together with the local nature of the TMS effects allows searching for compactly located groups of fibers.

Indeed, consistent with our hypothesis, the relations between observed and computed effects were revealed only for a small subset of the CST fibers, originating from the right hand-related area of the motor cortex. Note, that this area is located in the left hemisphere, since each hemisphere is controlling movements of the contralateral side of the body. Of importance, this group was stable across two independent sessions, suggesting that activation of these fibers is indeed responsible for the observed TMS effects.

Despite being statistically significant, the absolute correlation coefficients were not very high. This may be related to several factors. First, the stimulation coil used in the experimental sessions was not exactly the same as in the SimNIBS simulations, since SimNIBS provides only a limited number of the coil models. Although both coils had a figure-of-eight shape, this may have potentially influences the simulated field [17] and thus, the computed TMS effects. The second source of bias in the computed TMS effects may stem from errors

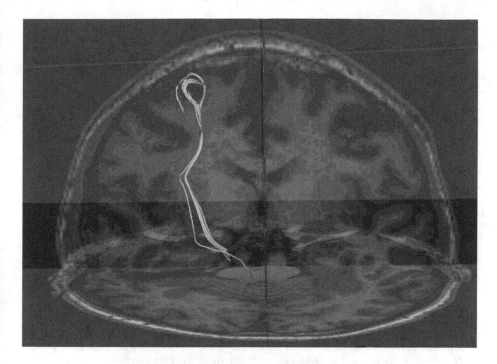

Fig. 4. The found group of CST streamlines.

in the estimation of the local fiber directions during tractography procedure. The applied tensor model is the simplest model that assumes only one principal diffusion direction within each image voxels. Other diffusion models, such as Q-ball imaging [2] or Constrained Spherical Deconvolution tractography [21], are believed to provide better estimation of the local fiber directions. However, there are no gold-standards on the tractography approaches and investigation of the influence of the tractography strategy upon the computed TMS effects was beyond the scope of the present work. Finally, the unexplained part of the TMS effects variability may be of physiological origin related to the undergoing non-stationary neuronal activity [14]. Nevertheless, even in the presence of the above mentioned limitations the proposed approach was able to reveal a relevant group of fibers which demonstrated a reliable relationship with the observed effects across independent experimental sessions.

5   Conclusion

In this paper we presented an approach for finding the group of streamlines related to the TMS-indused motor response. Each streamline corresponds to a set of nerve fibers thus making it possible to determine where the TMS-targeted nerve fibers are located. The found group was validated on a independent data, justifying the obtained result. It is also important to notice that our result suggests that closely related nerve fibers are responsible for similar behaviour.

Future work includes extension of the found results to other types of tractography with a potential in determination the best tractography approach. Moreover, this result should be validated on other subjects and different muscles. In particular, it is of high importance to determine whether the found streamlines are subject-related or their place is fixed, assuming that more precise atlases for navigation can be created.

Acknowledgement. Supported by Russian Science Foundation grant № 18-75-00034.

References

1. Atzmueller, M.: Subgroup discovery. Wiley Interdisc. Rew.: Data Min. Knowl. Discov. **5**(1), 35–49 (2015). https://doi.org/10.1002/widm.1144
2. Descoteaux, M., Angelino, E., Fitzgibbons, S., Deriche, R.: Regularized, fast, and robust analytical Q-ball imaging. Magn. Reson. Med. **58**(3), 497–510 (2007). https://doi.org/10.1002/mrm.21277
3. Devlin, J.T., Watkins, K.E.: Stimulating language: insights from TMS. Brain: J. Neurol. **130**(3), 610–622 (2007). https://doi.org/10.1093/brain/awl331
4. Garyfallidis, E., et al.: Dipy, a library for the analysis of diffusion MRI data. Front. Neuroinf. **8** (2014). https://doi.org/10.3389/fninf.2014.00008
5. Geeter, N.D., Crevecoeur, G., Leemans, A., Dupré, L.: Effective electric fields along realistic DTI-based neural trajectories for modelling the stimulation mechanisms of TMS. Phys. Med. Biol. **60**(2), 453–471 (2014). https://doi.org/10.1088/0031-9155/60/2/453
6. Hlustik, P., Solodkin, A., Gullapalli, R.P., Noll, D.C., Small, S.L.: Somatotopy in human primary motor and somatosensory hand representations revisited. Cerebral Cortex **11**(4), 312–321 (2001)
7. Iglesias, A.H.: Transcranial magnetic stimulation as treatment in multiple neurologic conditions. Curr. Neurol. Neurosci. Rep. **20**(1), 1–9 (2020). https://doi.org/10.1007/s11910-020-1021-0
8. Klomjai, W., Katz, R., Lackmy-Vallée, A.: Basic principles of transcranial magnetic stimulation (TMS) and repetitive TMS (rTMS). Ann. Phys. Rehabil. Med. **58**(4), 208–213 (2015). https://doi.org/10.1016/j.rehab.2015.05.005. Neuromodulation/Coordinated by Bernard Bussel, Djamel Ben Bensmail and Nicolas Roche
9. Knoch, D., Pascual-Leone, A., Meyer, K., Treyer, V., Fehr, E.: Diminishing reciprocal fairness by disrupting the right prefrontal cortex. Science **314**(5800), 829–832 (2006). https://doi.org/10.1126/science.1129156
10. Kulikova, S.: StimVis: a tool for interactive computation of the TMS-induced effects over tractography data. SoftwareX **12**, 100594 (2020). https://doi.org/10.1016/j.softx.2020.100594
11. Le Bihan, D., et al.: Diffusion tensor imaging: concepts and applications. J. Magn. Reson. Imaging **13**(4), 534–546 (2001). https://doi.org/10.1002/jmri.1076
12. Miranda, P.C., Correia, L., Salvador, R., Basser, P.J.: Tissue heterogeneity as a mechanism for localized neural stimulation by applied electric fields. Phys. Med. Biol. **52**(18), 5603–5617 (2007). https://doi.org/10.1088/0031-9155/52/18/009

13. Novikov, P., Nazarova, M., Nikulin, V.: TMSmap - software for quantitative analysis of TMS mapping results. Front. Hum. Neurosci. **12**(239) (2018). https://doi.org/10.3389/fnhum.2018.00239
14. Peters, J.C., Reithler, J., de Graaf, T.A., Schuhmann, T., Goebel, R., Sack, A.T.: Concurrent human TMS-EEG-fMRI enables monitoring of oscillatory brain state-dependent gating of cortico-subcortical network activity. Commun. Biol. **3**(40), 1176–1185 (2020)
15. Richter, L., Neumann, G., Oung, S., Schweikard, A., Trillenberg, P.: Optimal coil orientation for transcranial magnetic stimulation. PLoS One **8**(4) (2013). https://doi.org/10.1371/journal.pone.0060358
16. Roth, B.J., Basser, P.J.: A model of the stimulation of a nerve fiber by electromagnetic induction. IEEE Trans. Biomed. Eng. **37**(6), 588–597 (1990). https://doi.org/10.1109/10.55662
17. Salinas, F.S., Lancaster, J.L., Fox, P.T.: Detailed 3D models of the induced electric field of transcranial magnetic stimulation coils. Phys. Med. Biol. **52**(10), 2879–2892 (2007). https://doi.org/10.1088/0031-9155/52/10/016
18. Saturnino, G.B., Madsen, K.H., Thielscher, A.: Electric field simulations for transcranial brain stimulation using FEM: an efficient implementation and error analysis. J. Neural Eng. **16**(6), 066032 (2019). https://doi.org/10.1088/1741-2552/ab41ba
19. Silva, S., Basser, P.J., Miranda, P.C.: Elucidating the mechanisms and loci of neuronal excitation by transcranial magnetic stimulation using a finite element model of a cortical sulcus. Clin. Neurophys. **119**(10), 2405–2413 (2008). https://doi.org/10.1016/j.clinph.2008.07.248
20. Thielscher, A., Opitz, A., Windhoff, M.: Impact of the gyral geometry on the electric field induced by transcranial magnetic stimulation. Neuroimage **54**(1), 234–243 (2011). https://doi.org/10.1016/j.neuroimage.2010.07.061
21. Tournier, J.D., Calamante, F., Gadian, D.G., Connelly, A.: Direct estimation of the fiber orientation density function from diffusion-weighted MRI data using spherical deconvolution. Neuroimage **23**(3), 1176–1185 (2004). https://doi.org/10.1016/j.neuroimage.2004.07.037
22. Wagner, T.A., Zahn, M., Grodzinsky, A.J., Pascual-Leone, A.: Three-dimensional head model simulation of transcranial magnetic stimulation. IEEE Trans. Biomed. Eng. **51**(9), 1586–1598 (2004)
23. Windhoff, M., Opitz, A., Thielscher, A.: Electric field calculations in brain stimulation based on finite elements: an optimized processing pipeline for the generation and usage of accurate individual head models. Hum. Brain Mapp. **34**(4), 923–935 (2013)

Data Analysis in Astronomy

Data for Binary Stars from Gaia DR2

Dana Kovaleva[1]([✉]), Oleg Malkov[1], Sergei Sapozhnikov[1], Dmitry Chulkov[1],
and Nikolay Skvortsov[2]

[1] Institute of Astronomy, Moscow 119017, Russia
dana@inasan.ru
[2] Institute of Informatics Problems, Federal Research Center "Computer Science
and Control" of the Russian Academy of Sciences, Moscow 119333, Russia

Abstract. Gaia space mission has provided a large amount of astrometric and photometric data for the Milky Way stellar population. Binary
or multiple systems form a significant part of it. In spite of the fact that
Gaia DR2 treats all stars as singles in respect to astrometric solutions,
it still can be used to enrich our knowledge of the binary population
in several ways. Many stars known to be components of binaries have
received new astrometric measurements of validity to be discussed. Tens
of thousands of prospective binaries were discovered by different authors
as co-moving stars situated in space close to each other. We combine and
analyze the new information for binaries from Gaia DR2, and include it
into the Identification List of Binaries (ILB), a complete list of known
binary and multiple stars.

Keywords: Binary stars · Gaia data · Astronomical catalogues and
databases

1 Introduction

Binary and multiple stars are the significant part of the stellar population of
the Milky Way (see [5]). Various methods are used to discover and observe
binaries [17]. Whether certain binary star may be observed by a certain method
of observation, depends on its characteristics, both astrophysical (e.g. period,
semimajor axis, masses or evolutionary stage of the components) and geometrical
(distance, inclination of the orbit) (see discussion in [25]). The recovery of the
unbiased sample of binaries is important to understand star formation process.
Important steps regarding this problem were done in the recent decade with
the use of results of dedicated surveys of binaries of various observational types
(e.g. [2–4,13,20,21,27]), see also reviews [5,19]. The homogeneous parallaxes
provided by Hipparcos astrometric space mission [26] were used to choose the
targets of many surveys listed above. However, these surveys are small, limited
to hundreds of objects.

At the same time, there is more data on binary and multiple stars available
in catalogues and databases related to certain methods of observations. This

A. Sychev et al. (Eds.): DAMDID/RCDL 2020, CCIS 1427, pp. 125–133, 2021.
https://doi.org/10.1007/978-3-030-81200-3_9

data is non-homogeneous and difficult for data mining. Data in these datasets is often related to different categories of objects (components, pairs or systems), cross identification for the objects is often non-existent or unreliable. The Binary and multiple star DataBase (BDB) http://bdb.inasan.ru [10,15] uses the Identification List of Binaries, ILB [14,16], as a master catalogue providing cross-identification for all entities of binary and multiple stars. The purpose of BDB is to enable data mining for binaries from every catalogue or database.

Second data release of Gaia space mission [7] provided large amount of homogenous, high-accuracy astrometric and photometric data. In spite of the fact that Gaia DR2 treats all stars as singles in respect to astrometric solution, it still can be used to enrich our knowledge of binary population in several ways. Many stars known to be components of binaries have received new astrometric parameters, some of these binaries are resolved [28]. Tens of thousands of prospective binaries were discovered by different authors [6,9,22,23] as co-moving stars situated in space close to each other. Some of these new systems overlap with known ones.

The purpose of the current research is to compare information on binary and multiple stars before and after Gaia DR2, to identify objects in the Gaia DR2 catalogue related to binary systems, and to incorporate this data into ILB catalogue and BDB database. We also discuss how binary stars are represented in the Gaia DR2.

Section 2 describes cross-identification of Gaia DR2 with the catalogues of binary stars (ILB and Catalogue of visual binary stars with orbits ORB6 [8]). Section 3 refers to catalogues of co-moving stars based on Gaia DR2, and its relations with previously known binary and multiple stars. In Sect. 4 we discuss the results, and Sect. 5 contains our conclusions.

2   Catalogues of Binary Stars and Its Cross Identification with Gaia DR2

The data on components, binary and multiple systems is basically cross-identified, assigned with unique identifiers [11] and stored in ILB catalogue. Significant part of ILB items refers to visual binaries from The Washington Visual Double Star Catalog [18]. Large part of them are pairs with slow or non-registerable relative motion. For these stars cross-identification process is quite straightforward. Situation may be different for stars with significant orbital motion.

2.1   ILB

Originally ILB was constructed to include all possible entities from the catalogues of binary and multiple stars. The recent version of ILB contains information on 554820 entities (191755 pairs and 363065 components) within 165221 binary and multiple systems (note that a pair may consist of two components, two pairs or one component and one pair, while a component always matches one star). Every record for component in ILB contains coordinates from source catalogue. We have performed cross-identification with Gaia DR2 source catalogue addressing via Topcat client to the Gaia-TAP interface supported by ARI http://gaia.ari.uni-heidelberg.de/tap.html, with matching radius set to 1″. We consider only cross-identification of components. We label three modes of identification based on the association of components with Gaia DR2 sources. 56399 pairs are "separated", which means that for each component a separate unique Gaia DR2 source was successfully identified. 33649 pairs are "joint": both components are identified, but are associated with the same source in Gaia DR2 (which means that the binary is unresolved by Gaia). Finally, for some pairs only one of the components was identified as a Gaia DR2 source, and no corresponding source was found for another component. These latter identifications are referred to as "singles". A total of 100933 "single" components of different pairs is present.

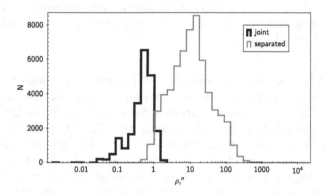

Fig. 1. ILB pairs cross-identified with Gaia DR2: distribution over angular distances between components. Black histogram ("joint") – pairs unresolved by Gaia, both components having identification with one Gaia DR2 source, grey histogram ("separated") – pairs where each component is identified with separate Gaia DR2 source.

Figure 1 represents distribution over angular distances from ILB between components for the pairs identified with Gaia DR2 as one source ("joint") and as two separate sources ("separated"). Note that ILB may include pairs that are not physical (often, from the WDS), so we do not discuss here upper limit of this separation.

Figure 2 represents distribution of "separated" dataset over magnitude difference ΔG (Gaia Gmag) as a function of angular separation between the components ρ calculated from ILB coordinates. See discussion of these results in Sect. 4.

2.2   ORB6

The Catalogue of visual binary stars with orbits ORB6 [8] contains data on orbital solutions for visual and astrometric binaries. The catalogue is regularly

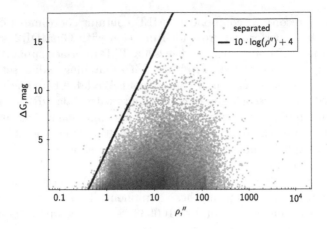

Fig. 2. ILB pairs cross-identified with two separate Gaia DR2 sources: Gaia G magnitude difference between the components vs angular separation between the components. The black line is approximate limiting resolution depending on magnitude difference, see discussion in Sect. 4.

updated; as for June 2019, it contained 2942 orbital solutions for 2858 systems. This catalogue includes visual and astrometric binaries with known orbital motion. Since Gaia DR2 does not account for orbital motion of the binary [7], one should be careful studying such systems with Gaia. Expected lower resolution limit of Gaia DR2 is estimated as ≈0.4", and it is expected that almost all pairs with separation larger than 2.2" are resolved [7]. Ephemerides provided in ORB6 predict angular separation between components larger than these values for 27% and 6% of pairs, respectively. We have been searching for Gaia DR2 sources around the coordinates from ORB6 [1]. Primary radius of search was selected as 3" since at larger distances number of objects within the field critically increases. If the selected field contained only one Gaia DR2 source, we have compared its stellar magnitude G with the one indicated in ORB6; the identification was considered successful if the difference was less than 1 magnitude. If the examined field contained two Gaia DR2 sources, we have calculated angular separation and positional angle between stars and compared them with the ephemerides by ORB6 for the J2016.0 epoch (Gaia DR2 coordinates are for J2015.5 epoch). Some pairs of ORB6 belong to systems of higher multiplicity, the data of WDS catalogue was used to distinguish between components. At first stage, about 65% of ORB6 pairs received at least one identification with Gaia DR2 source. For initially non-identified objects of ORB6 we slightly relaxed cross-matching conditions. Since most ORB6 stars are bright, contamination with the field stars in Gaia is low, the largest obstacle is the correct processing of multiple systems. As a result, we found at least one identification with Gaia DR2 sources for 88% of ORB6 pairs.

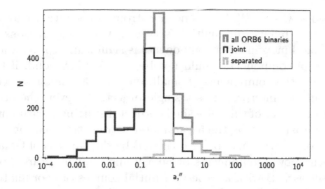

Fig. 3. ORB6 pairs cross-identified with Gaia DR2: distribution over angular distances between components. Grey histogram – all ORB6 stars, black histogram ("joint") – pairs which have identification with one Gaia DR2 source (not necessarily referring to both components), grey histogram ("separated") – pairs where each component is identified with separate Gaia DR2 source.

Figure 3 represents distribution over angular distances between components for all ORB6 pairs, pairs identified with Gaia DR2 as one source ("joint") or two separate sources ("separated").

3   Catalogues of Co-moving Stars from Gaia DR2

Publication of Gaia DR2 have stimulated several groups to independently start a search for stars close in 5-dimensional space of Gaia parameters (coordinates, proper motions, and parallaxes). These groups used different filters to discover co-moving pairs considered as binary candidates; thus the datasets overlap but are not identical. We considered three catalogues: the dataset by [6] includes 55507 binary stars within 200 pc, [9] contains 3741 binary and multiple stars with bright components included into the Tycho-2 catalogue, and [22] with 9977 pairs within 100 pc. All listed catalogues use quality cuts recommended for astrometrically and photometrically "pure" Gaia DR2 data ([12], and https://www.cosmos.esa.int/web/gaia/dr2-known-issues) to decrease number of false binaries candidates. Different reasonable limitations for maximum distance between the components, parallax and flux errors are used.

Cross-identification between these three catalogues and ILB (made by the unique Gaia DR2 identifier of components) leaves 62980 pairs, 49638 of them are previously unknown as binaries. Data on these stars is included into ILB as the result of this research.

4   Discussion

There is a number of reasons why certain binary and multiple stars may be not identified in Gaia DR2. Among them is the incompleteness of Gaia for fast-moving stars, bright stars and objects around them, as well as in dense stellar fields [7]. Besides, Gaia DR2 5-parameter astrometric solution may be unsatisfactory for stars in binary and multiple systems. A couple of reasons is possible. First, if the orbital movement of components is significant, the solution ignoring mutual motion of components would not be successful. Second, if the angular distance between the components is small and they are not resolved by Gaia, the binary would be interpreted as a single object. However, the orbital movement shifts photocenter of this false single object distorting measurements of the coordinates approximated by the 5-parameter astrometric solution.

The effects of binarity may be investigated by the analysis of Gaia DR2 data associated with known components of binary and multiple stars. For the compilative dataset as ILB with a variety of initial sources of coordinates, selected radius of matching of $1''$ may be insufficient, while increase of matching radius results in increase of a number of cross-identifications, inevitably increasing risk of false identification.

4.1   Astrometric Solutions for Binary Stars

Next we consider quality markers (RUWE and Flux excess factor) of the astrometric solutions for known binary stars discussed in Sect. 2. We need some reference dataset for comparison, Hipparcos stars cross-identified with Gaia DR2 source catalogue in agreement with "Best neighbour" table Hipparcos-Gaia DR2 are used. Again, we use Topcat client to access Gaia data via TAP interface at ARI. For the resulting 83034 Hipparcos stars 7% of dataset do not pass quality cut on RUWE, and just 2% do not pass quality cut on Flux excess factor.

Table 1. Quality of astrometric solutions for Gaia DR2 sources identified with binary stars.

Dataset	G_{max}	$\log \varpi_{max}$	R	F	RF
Hipparcos	8.5	3.5	7	2	8
ILB-joint	10.5	3.5	40	15	45
ILB-separated (A)	11	3.0	16	11	22
ILB-separated (B)	12	2.0	15	17	26
ORB6-joint	8.0	17.0	69	17	77
ORB6-separated (A)	8.2	20.0	29	36	47
ORB6-separated (B)	9.0	20.0	39	61	67

Table 1 contains these fractions for different binary datasets. Columns meaning is as follows: **Dataset**—dataset short name (Hipparcos – reference dataset, ILB-joint – binary stars from ILB catalogue associated with one Gaia DR2 source; ILB-separate (A) – binary stars from ILB catalogue associated with two separate Gaia DR2 source, main components; ILB-separate (B) – binary stars from ILB catalogue associated with two separate Gaia DR2 source, secondary components; lines with ORB6 as prefix – similar data for binary stars from ORB6 catalogue; G_{max}, **mag**—maximum of Gaia G magnitude distribution for this dataset; $\log \varpi_{max}$,"—maximum of log of parallax distribution (in mas) for this dataset (approximately); **R(RUWE > 1.4, %)**—percentage of stars failed to pass quality cut on RUWE; **F (F.e.f cut failed, %)**—percentage of stars failed to pass quality cut on Flux excess factor; **RF (RUWE or F.e.f., %)**—percentage of stars failed to pass at least one of quality cuts.

Components of binaries projected close to each other have a significant probability to fail quality cuts. This probability is large for unresolved binaries, and for resolved pairs it is larger for secondary component than for primary one. For more distant stars quality markers become "better", which is natural (orbital movement becomes less and less noticeable as distance increases). On the other hand, binary star failing or passing quality cuts evidently does not depend on its apparent magnitude.

4.2 ILB Update with Gaia DR2 Data and Gaia DR2 Binaries

As a result of described work, the catalogue of identifications of binaries ILB has been enhanced with data from Gaia DR2 catalogue: 49638 new components, with corresponding items for pairs and systems, are added; Gaia DR2 identifiers and data are added to both components of 56399 pairs ("separated" identifications), to a single component in 100933 pairs ("single" identifications), and to a pair as a whole entity in 33649 cases ("joint" identifications). The updated ILB will be available in VizieR and implemented into the BDB database.

5 Results and Conclusions

We have considered how data for known binary stars assembled in the ILB catalogue is represented in the Gaia DR2 source catalogue and made cross-identifications when possible. We have cross-matched the datasets of co-moving candidates for binaries discovered in Gaia DR2 with each other and ILB, compiled the new updated version of ILB enhanced with new data and new records from Gaia DR2.

It was investigated how markers of quality of astrometric and photometric solution behave for the data on known binary stars in Gaia DR2. Resolved components of binary stars exhibiting visible orbital motion within about 50 pc, and unresolved binaries at distances to 200–300 pc fail quality cuts recommended by Gaia-ESA in 45%–80% of cases.

Acknowledgement. The study was partly funded by RFBR, project numbers 19-07-01198, 18-07-01434. The use of TOPCAT tool http://www.starlink.ac.uk/topcat/ [24] is gratefully acknowledged. This work has made use of data from the European Space Agency (ESA) mission *Gaia* (https://www.cosmos.esa.int/gaia), processed by the *Gaia* Data Processing and Analysis Consortium (DPAC, https://www.cosmos.esa.int/web/gaia/dpac/consortium). Funding for the DPAC has been provided by national institutions, in particular the institutions participating in the *Gaia* Multilateral Agreement. This research has made use of the VizieR catalogue access tool, CDS, Strasbourg, France (DOI: 10.26093/cds/vizier). The original description of the VizieR service was published in 2000, A&AS 143, 23.

References

1. Chulkov D.A.: Objects of the catalog of orbits of visual binary stars in Gaia DR2. In: Proceedings of INASAN, vol. 3, pp. 360–365 (2019). (in Russian), https://doi.org/10.26087/INASAN.2019.3.1.056. http://www.inasan.ru/wp-content/uploads/2019/11/ntr03.pdf#page=360
2. Cortés-Contreras, M., et al.: CARMENES input catalogue of M dwarfs. II. High-resolution imaging with FastCam. Astron. Astrophys. **597**, A47 (2017). https://doi.org/10.1051/0004-6361/201629056
3. De Rosa, R.J., et al.: The VAST survey - III. The multiplicity of A-type stars within 75 pc. Mon. Not. R. Astron. Soc. **437**(2), 1216–1240 (2014). https://doi.org/10.1093/mnras/stt1932
4. Dieterich, S.B., Henry, T.J., Golimowski, D.A., Krist, J.E., Tanner, A.M.: The solar neighborhood. XXVIII. The multiplicity fraction of nearby stars from 5 to 70 AU and the brown dwarf desert around M dwarfs. Astron. J. **144**(2), 64 (2012). https://doi.org/10.1088/0004-6256/144/2/64
5. Duchêne, G., Kraus, A.: Stellar multiplicity. Ann. Rev. Astron. Astrophys. **51**, 269–310 (2013). https://doi.org/10.1146/annurev-astro-081710-102602
6. El-Badry, K., Rix, H.W.: Imprints of white dwarf recoil in the separation distribution of Gaia wide binaries. Mon. Not. R. Astron. Soc. **480**(4), 4884–4902 (2018). https://doi.org/10.1093/mnras/sty2186
7. Gaia Collaboration, Brown, A.G.A., Vallenari, A., Prusti, T., et al.: Gaia data release 2. Summary of the contents and survey properties. Astron. Astrophys. **616** A1 (2018). https://doi.org/10.1051/0004-6361/201833051
8. Hartkopf, W.I., Mason, B.D., Worley, C.E.: The 2001 US naval observatory double star CD-ROM. II. The fifth catalog of orbits of visual binary stars. Astron. J. **122**(6), 3472–3479 (2001). https://doi.org/10.1086/323921
9. Jiménez-Esteban, F.M., Solano, E., Rodrigo, C.: A catalog of wide binary and multiple systems of bright stars from Gaia-DR2 and the virtual observatory. Astron. J. **157**(2), 78 (2019). https://doi.org/10.3847/1538-3881/aafacc
10. Kovaleva, D., Kaygorodov, P., Malkov, O., Debray, B., Oblak, E.: Binary star DataBase BDB development: structure, algorithms, and VO standards implementation. Astron. Comput. **11**, 119–127 (2015). https://doi.org/10.1016/j.ascom.2015.02.007
11. Kovaleva, D.A., Malkov, O.Y., Kaygorodov, P.V., Karchevsky, A.V., Samus, N.N.: BSDB: a new consistent designation scheme for identifying objects in binary and multiple stars. Baltic Astron. **24**, 185–193 (2015)
12. Lindegren, L., et al.: Gaia data release 2 The astrometric solution. Astron. Astrophys. **616**, A2 (2018). https://doi.org/10.1051/0004-6361/201832727
13. Malkov, O., et al.: Evaluation of binary star formation models using well-observed visual binaries. In: Manolopoulos, Y., Stupnikov, S. (eds.) Data Analytics and Management in Data Intensive Domains, Proc. XX International Conference, DAMDID/RCDL 2018, Moscow, Russia, Oct 2018, Revised Selected Papers, Communications in Computer and Information Science (CCIS), vol. 1003, pp. 91–107. Springer International Publishing (2019)

14. Malkov, O., Karchevsky, A., Kaygorodov, P., Kovaleva, D.: Identification list of binaries. Baltic Astron. **25**, 49–52 (2016)
15. Malkov, O., Karchevsky, A., Kaygorodov, P., Kovaleva, D., Skvortsov, N.: Binary star database (BDB): new developments and applications. Data **3**(4) (2018). https://doi.org/10.3390/data3040039. https://www.mdpi.com/2306-5729/3/4/39
16. Malkov, O., Karchevsky, A., Kovaleva, D., Kaygorodov, P., Skvortsov, N.: Catalogue of identifications of objects in binary and multiple stars. Astronomichesky Zhurnal Suppl. Ser. **95**(7), 3–16 (2018). (in Russian), https://doi.org/10.1134/s0004629918070058
17. Malkov, O., Kovaleva, D., Kaygorodov, P.: Observational types of binaries in the binary star database. In: Balega, Y.Y., Kudryavtsev, D.O., Romanyuk, I.I., Yakunin, I.A. (eds.) Stars: From Collapse to Collapse, Astronomical Society of the Pacific Conference Series, vol. 510, p. 360 (2017)
18. Mason, B.D., Wycoff, G.L., Hartkopf, W.I., Douglass, G.G., Worley, C.E.: VizieR Online Data Catalog: The Washington Visual Double Star Catalog (Mason+ 2001–2014). VizieR Online Data Catalog, vol. 1 (2016)
19. Moe, M., Di Stefano, R.: Mind your Ps and Qs: the interrelation between period (P) and mass-ratio (Q) distributions of binary stars. Astrophys. J. Suppl. Ser. **230**, 15 (2017). https://doi.org/10.3847/1538-4365/aa6fb6
20. Raghavan, D., et al.: A survey of stellar families: multiplicity of solar-type stars. Astrophys. J. Suppl. Ser. **190**(1), 1–42 (2010). https://doi.org/10.1088/0067-0049/190/1/1
21. Sana, H., et al.: Southern massive stars at high angular resolution: observational campaign and companion detection. Astrophys. J. Suppl. Ser. **215**(1), 15 (2014). https://doi.org/10.1088/0067-0049/215/1/15
22. Sapozhnikov S.A., Kovaleva D.A., Malkov O.Yu.: Catalogue of visual binaries in Gaia DR2. In: Proceedings of INASAN, vol. 3, pp. 366–371 (2019). (in Russian), https://doi.org/10.26087/INASAN.2019.3.1.057. http://www.inasan.ru/wp-content/uploads/2019/11/ntr03.pdf#page=366
23. Sapozhnikov, S.A., Kovaleva, D.A., Malkov, O.Yu., Sytov, A.Yu.: Population of binaries with common proper motions in Gaia DR2. Astronomichesky Zhurnal **97**(10), 1–14 (2020). (in Russian)
24. Taylor, M.B.: TOPCAT & STIL: Starlink table/votable processing software. In: Shopbell, P., Britton, M., Ebert, R. (eds.) Astronomical Data Analysis Software and Systems XIV ASP Conference Series, Vol. 347, Proceedings of the Conference held 24–27 October, 2004 in Pasadena, California, USA. Edited by P. Shopbell, M. Britton, and R. Ebert. San Francisco: Astronomical Society of the Pacific, 2005, p. 29, Astronomical Society of the Pacific Conference Series, vol. 347, p. 29 (2005)
25. Tokovinin, A.: From binaries to multiples. I. data on F and G dwarfs within 67 pc of the Sun. Astron. J. **147**, 86 (2014). https://doi.org/10.1088/0004-6256/147/4/86
26. van Leeuwen, F.: Validation of the new Hipparcos reduction. Astron. Astrophys. **474**(2), 653–664 (2007). https://doi.org/10.1051/0004-6361:20078357
27. Ward-Duong, K., et al.: The M-dwarfs in multiples (MINMS) survey - I. Stellar multiplicity among low-mass stars within 15 pc. Mon. Not. R. Astron. Soc. **449**(3), 2618–2637 (2015). https://doi.org/10.1093/mnras/stv384
28. Ziegler, C., et al.: Measuring the recoverability of close binaries in Gaia DR2 with the Robo-AO Kepler survey. Astron. J. **156**(6), 259 (2018). https://doi.org/10.3847/1538-3881/aad80a

Classification Problem and Parameter Estimating of Gamma-Ray Bursts

Pavel Minaev[1,2]([⊠]) [iD] and Alexei Pozanenko[1,2,3]

[1] Space Research Institute (IKI), 84/32 Profsoyuznaya Str., 117997 Moscow, Russia
[2] Moscow Institute of Physics and Technology (MIPT), Institutskiy Pereulok, 9, 141701 Dolgoprudny, Russia
[3] National Research University Higher School of Economics, 101000 Moscow, Russia

Abstract. There are at least two distinct classes of Gamma-Ray Bursts (GRB) according to their progenitors: short duration and long duration bursts. It was shown that short bursts result from compact binary merging, while long bursts are associated with core collapse supernova. However, one could suspect the existence of more classes and subclasses. For example, compact binary can be double neutron stars, or neutron star and black hole, which might generate gamma-ray bursts with different properties. From another hand, gamma-ray transients are known to be produced by magnetars in Galaxy, named Soft Gamma-Repeaters (SGR). A Giant Flare from SGR can be detected from a nearby galaxy, and it can mimic for a short GRB. So the classification problem is very important for correct investigation of different transient progenitors. Gamma-ray transients are characterized by a number of parameters and known phenomenology correlations between them, obtained for well classified ones. Using these correlations, which could be unique for different classes of gamma-ray transients, we can classify an event and determine the type of its progenitor, using only temporal and spectral characteristics. We suggest the statistical classification method, based on the cluster analysis of the trained dataset of 323 events. Using the known dependencies, one can not only classify the types of gamma-ray bursts, but also discriminate events that are not associated with gamma-ray bursts, but have a different physical nature. We show that GRB 200415A, originally classified as a short GRB, probably does not belong to the class of short GRBs, but it is most likely associated with the giant flare of SGR. On the other hand, we can estimate one of the unknown parameters if we assume the certain classification of the event. As an example, an estimate the redshift of the GRB 200422A source is given. We also discuss that in some cases it is possible to give a probabilistic estimate of the unknown parameters of the source. The method could be applied to any other analogous classification problems.

Keywords: Gamma-ray burst · Soft gamma repeater · Statistics · Correlation · Classification · Cluster analysis · Redshift

© Springer Nature Switzerland AG 2021
A. Sychev et al. (Eds.): DAMDID/RCDL 2020, CCIS 1427, pp. 134–147, 2021.
https://doi.org/10.1007/978-3-030-81200-3_10

1   Introduction

Two classes of gamma-ray bursts (GRB) were discovered in a series of KONUS experiments [27] and confirmed in BATSE experiment [21]. Type I bursts, in general, are characterized by a short duration ($T_{90} < 2\,$s), a harder energy spectrum, less distinct spectral evolution (lag) [10,21,32,35,37,39]. Duration and hardness ratio distributions, used for the blind classification of GRBs traditionally, are overlapped significantly, making the problem of the blind classification still actual (especially in the overlap region) [33,36]. The correct classification is crucial for understanding GRB progenitors.

Type I bursts are associated with a merger of compact binaries, consisted of two neutron stars [6,29,42], which was recently confirmed by the gravitational wave experiments LIGO/Virgo for GRB 170817A [2,3,46]. Several type I bursts are also accompanied by an additional component with a duration of tens of seconds and a soft energy spectrum – the extended emission (EE) [9,17,35,36, 40], which nature is still unclear [5,30,35,49].

Type II bursts are associated with a core collapse of supermassive stars [28,43,54], some of them are also accompanied by a bright (intrinsically) Ic supernovae [7,16,43,53].

The another source of hard and short gamma-ray emission is a soft gamma repeater (SGR) during the giant flare (GF), which could be observed from our or nearby galaxies [14,26]. The light curve and spectral characteristics of the main pulse of the giant flare are similar to ones of type I GRBs, complicating a blind classification of short bursts in case of the absence of precise localization of the source [14,26]. SGRs are usually associated with magnetars, highly magnetized single neutron stars, while the nature of giant flares remains unresolved [22].

Gamma-ray bursts are characterized by a number of correlations between different observational parameters. In this work we investigate the $E_{p,i}$ – E_{iso} correlation [4], the correlation between spectral hardness $E_{p,i}$ and total energetics E_{iso} of the prompt emission. One of possible explanations of the $E_{p,i}$ – E_{iso} correlation is connected with viewing angle effects: bursts observed close to the axis of relativistic jet are brighter (larger E_{iso}) and harder (larger $E_{p,i}$), while the bursts seen off-axis are fainter and softer [11,23,46].

Recently, it was shown, that the $E_{p,i}$ – E_{iso} correlation could be used as the classification indicator of GRBs: the correlation for type I bursts is found to be well-distinguished from the one constructed for type II bursts [24,34,46,48,57]. In [34] we compiled the most extensive GRB sample and confirmed the possibility of using the correlation to independently classify GRBs. The correlation was found to have a similar power-law index value for both type I and type II bursts, $E_{p,i} \sim E_{iso}^{0.4}$, which possibly indicates the same emission mechanism of both GRB types. In [34] we also suggested the new classification method, based on the $E_{p,i}$ – E_{iso} correlation and introduced two parameters, EH and EHD. The EHD parameter was found to be the most reliable for the blind type I/type II classification. The method has already been applied to restrict a jet opening angle for GRB 190425, associated with the merging of binary neutron stars [47] registered by LIGO/Virgo [1].

In this work, we continue the statistical investigation of the $E_{p,i} - E_{iso}$ correlation, including the cluster analysis, using the sample of 320 GRBs and 3 SGR GFs as trained dataset. We demonstrate the possibility of classifying GRBs and estimating their redshift on testing sample of GRB 200415A and GRB 200422A, which type and redshift are unknown.

2   The $E_{p,i} - E_{iso}$ Correlation

2.1   Constructing and Fitting the Correlation

The parameter $E_{p,i}$ characterizes the hardness of the energy spectrum and represents the position of the extremum (maximum) of the prompt emission energy spectrum νF_ν in the rest frame, calculated as $E_{p,i} = E_p(1 + z)$, where E_p is the observed value.

E_{iso} is the isotropic equivalent total energy, emitted in gamma-rays in (1, 10000) keV range, calculated as $E_{iso} = \frac{4\pi D_L^2 F}{1+z}$, where F is the burst fluence in (1, 10000) keV energy range in the rest frame and D_L is the luminosity distance calculated using standard cosmological parameters, $H_0 = 67.3\,\mathrm{km\,s^{-1}\,Mpc^{-1}}$, $\Omega_M = 0.315$, $\Omega_\Lambda = 0.685$ [44].

$$\lg\left(\frac{E_{p,i}}{100\ keV}\right) = a \lg\left(\frac{E_{iso}}{10^{51}\ erg}\right) + b. \tag{1}$$

To investigate the $E_{p,i} - E_{iso}$ correlation (Eq. (1)) we use the sample consisting of 45 type I and 275 type II bursts with corresponding approximation results from our previous work [34]. This sample contains bursts with reliable classification and observational parameters (including redshift) and represents our trained dataset. The sample could be expanded, as new bursts with reliable characteristics will be registered.

We also include in our analysis the giant flare (GF) of soft gamma repeater SGR 1806-20 on December 27, 2004 [14], and two SGR GF extragalactic candidates from M31 [26] and M81/M82 [15], shown by open square symbols at Fig. 1.

Table 1. The $E_{p,i} - E_{iso}$ correlation fit parameters.

Sample	N	a	b	σ_{cor}
I	45	0.41 ± 0.05	0.83 ± 0.06	0.32
II	275	0.43 ± 0.03	-0.24 ± 0.06	0.29

The $E_{p,i} - E_{iso}$ diagram constructed for this sample is presented at Fig. 1, fit parameters are listed in Table 1.

2.2 Using the Correlation to Classify GRBs, *EH* Parameter

As seen on Fig. 1, samples of type I bursts, type II bursts and SGR giant flares are well separated, giving the possibility to use the correlation to classify these transients.

Power-law indexes of the $E_{p,i} - E_{iso}$ correlation are the same (within 1σ) for both type I and type II bursts, $a \simeq 0.4$, while b values differ significantly: the correlation for type I bursts is shifted up against the one for type II bursts.

The positions of two SGR GF candidates on the $E_{p,i} - E_{iso}$ plane are close to the confirmed one of SGR 1806-20, supporting the association of these candidates with the class of SGR giant flares. Moreover, the positions of all these three events do not contradict the existence of the $E_{p,i} - E_{iso}$ correlation for SGR GF with $a \simeq 0.4$.

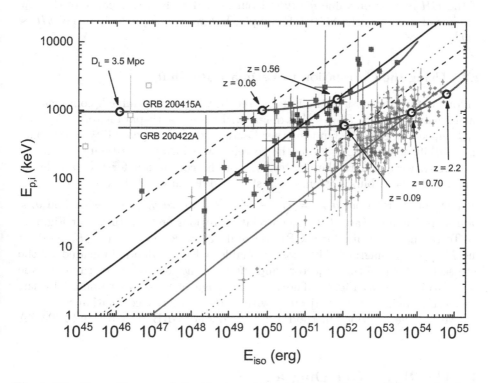

Fig. 1. The $E_{p,i} - E_{iso}$ correlation for type I bursts (squares), type II bursts (circles) and SGR giant flares (open squares) with corresponding fits. $2\sigma_{cor}$ correlation regions are also presented for type I (type II) bursts by dashed (dotted) lines. Trajectories (dependencies on redshift) for GRB 200415A and GRB 200422A are shown by lines with key points indicated by open circles (see text). (Color figure online)

Described features of the correlation can be used in a blind classification of the GRB-like events: the distribution of deviations from the power-law dependence $E_{p,i} \sim E_{iso}^{0.4}$ should be trimodal (type I GRBs, type II GRBs, SGR GF).

The deviation could be measured by the parameter EH ("Energy-Hardness") –
the combination of $E_{p,i}$ and E_{iso} parameters (Eq. 2).

$$EH = \frac{(E_{p,i}/100 \ keV)}{(E_{iso}/10^{51} \ erg)^{\ 0.4}}. \tag{2}$$

The separation value $EH = 3.3$, which can be used in a blind classification
of gamma-ray bursts, was found in [34] by investigating the EH distribution
for the sample of 320 GRBs. A burst with $EH > 3.3$ should be classified as the
type I one. Details of the classification method, including reliability tests and
comparison with other GRB classification schemes are presented in [34].

For all three SGR giant flares we obtain $EH > 600$, which is 20 times larger,
than the value of any type I burst from our sample. We omit the estimation
of the EH separation value for type I bursts and SGR GF, because of the low
number of events in the SGR GF sample, but we can suppose it to be $EH \sim$
100.

2.3   Using the Correlation to Estimate Redshift

Unfortunately, for most GRBs (>90%) redshift is not determined, complicating
their analysis on the $E_{p,i}$ – E_{iso} diagram, because both parameters, $E_{p,i}$ and
E_{iso} depend on redshift. On the other hand, it gives us a possibility to estimate
the redshift of such events using the correlation, which is investigated below.

Firstly, we calculate the scatter σ_{cor} of the correlation for the samples of
type I and type II bursts as empirical standard deviation of the events positions
against the correlation: $\sigma_{cor} = \sqrt{\frac{\sum_{i=1}^{n}(y_i - \bar{y}_i)^2}{n-1}}$, where $\bar{y}_i = ax_i + b$. The σ_{cor}
values are listed in Table 1, $2\sigma_{cor}$ correlation regions are also shown at Fig. 1.

To estimate redshift for a GRB we build a trajectory $(E_{p,i}(z), E_{iso}(z))$ on
the $E_{p,i}$ – E_{iso} diagram. The most probable redshift value is indicated by the
intersection point of the trajectory and the fit of the corresponding sample (blue
or red thick line at Fig. 1). The possible range of redshift is limited by the
intersection points of the trajectory with borders of $2\sigma_{cor}$ correlation region.

We test the described method using the data of GRB 200415A and 200422A
in Sect. 4.2 and 5.2.

3   The $T_{90,i}$ – EH Diagram

In this section we consider another possibility of using the $E_{p,i}$ – E_{iso} correlation
to classify gamma-ray bursts and to estimate their redshift, by taking into a
consideration a duration of the bursts additionally.

3.1   Using the Diagram to Classify GRBs, EHD Parameter

The duration distribution of GRBs is bimodal: type I bursts are shorter than
type II ones with separation value of $T_{90} \sim 2$ s [21,35]. The T_{90} parameter is

the time interval with integrated counts raise from 5% to 95% [20]. Although T_{90} value is highly affected by selection effects, it can be a good additional classification indicator. The observed duration depends on redshift of the source $T_{90} = T_{90,i}(1 + z)$, where $T_{90,i}$ is the intrinsic (rest-frame) duration. Therefore, in our analysis we use the intrinsic value $T_{90,i}$.

The $T_{90,i} - EH$ diagram is presented at Fig. 2. Type I bursts are placed at the top left on the diagram being short hard and faint, comparing with type II bursts. SGR giant flares are placed at very top left on the diagram, having typical duration of type I bursts, $T_{90,i} \sim 0.15\,\mathrm{s}$, but sufficiently higher value of the EH parameter.

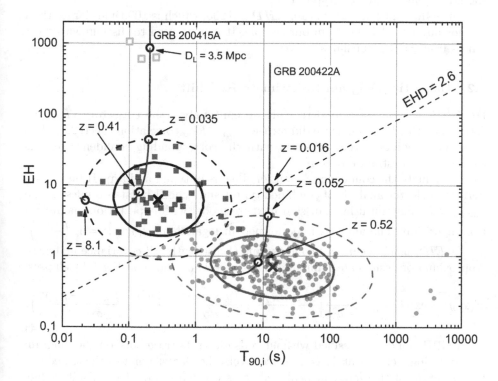

Fig. 2. The $T_{90,i} - EH$ diagram for type I bursts (squares), type II bursts (circles) and SGR giant flares (open squares). The dashed line ($EHD = 2.6$) is used in a blind classification of GRBs. Gaussian mixture model fits for type I and type II bursts are shown by thick solid and dashed ellipses, representing 1σ and 2σ confidence regions, correspondingly. Trajectories (dependencies on redshift) for GRB 200415A and GRB 200422A are shown by black lines with key points indicated by open circles (see text). (Color figure online)

The separation of the clusters is not satisfactory, when using only $T_{90,i}$ (separation, parallel to y-axis) or EH (separation, parallel to x-axis) parameters. Good separation is reached with the power-law dependence, $EH \sim T_{90,i}^{0.5}$. Therefore, we can modify the EH parameter by including the duration $T_{90,i}$ into a consideration and obtain the second classification parameter, EHD ("Energy-Hardness-Duration", Eq. 3).

$$EHD = \frac{EH}{(T_{90,i}/1\ s)^{0.5}} = \frac{(E_{p,i}/100\ keV)}{(E_{iso}/10^{51}\ erg)^{0.4}\ (T_{90,i}/1\ s)^{0.5}}. \tag{3}$$

The EHD parameter results in the best separation of type I bursts from type II ones, with the separation value of $EHD = 2.6$ (see details in [34]). As a consequence, it gives the best reliability in a blind classification. If the value of EHD parameter is greater than the separation value $EHD = 2.6$, the burst should be classified as the type I one.

For SGR giant flares we obtain $EHD > 1200$, which is 100 times larger than one for any type I burst from our sample. It can be used to distinguish type I bursts from SGR giant flares.

3.2   Using the Diagram to Estimate Redshift

For GRBs without measured redshift we can analyze the trajectory $(T_{90,i}(z),$ $EH(z))$ on the diagram, as we did for the $E_{p,i} - E_{iso}$ correlation (Sect. 2.3). The intersection points of the trajectory with the corresponding confidence regions can be used to estimate redshift.

To evaluate the confidence regions of clusters of type I and type II bursts, we perform a cluster analysis of combined type I and type II bursts samples on the $T_{90,i} - EH$ diagram using gaussian mixture model with expectation maximization algorithm. The analysis is performed in logarithmic space $x = \lg\left(\frac{T_{90,i}}{1s}\right)$, $y = \lg(EH)$, using a sum of two gaussians (Eq. 4) with full covariance matrices. Approximation results are summarized in Table 2 and shown at Fig. 2 by ellipses.

$$F(x,y) = \exp\left(-\frac{1}{2(1-\rho^2)}\left[\frac{(x-\mu_X)^2}{\sigma_X^2} + \frac{(y-\mu_Y)^2}{\sigma_Y^2} - \frac{2\rho(x-\mu_X)(y-\mu_Y)}{\sigma_X\sigma_Y}\right]\right). \tag{4}$$

A GRB could be classified without redshift, if its trajectory on the diagram crosses a single cluster at 2σ confidence level. The closest point of the trajectory to a position of the gaussian peak indicates the most probable redshift value, while intersection points of the trajectory with 2σ confidence regions limit possible redshift interval. Using confidence regions of clusters instead of the EHD parameter value sufficiently improves the reliability of the $T_{90,i} - EH$ diagram to estimate redshift.

Table 2. The $T_{90,i} - EH$ diagram fit parameters.

Sample	μ_X	μ_Y	σ_X^2	σ_Y^2	ρ
I	−0.57	0.80	0.20	0.12	−0.063
II	1.14	−0.15	0.37	0.084	−0.18

We estimate redshift for GRB 200415A and 200422A, using the $T_{90,i} - EH$ diagram, in Sect. 4.3 and 5.3.

4   GRB 200415A

4.1   Observations

The short duration ($T_{90} \simeq 0.2\,\text{s}$) GRB 200415A was registered at 08:48:05 UT on 15 Apr 2020 by a number of gamma-ray experiments [12,13,25,41,45]. It allowed to localize the burst using triangulation method [50,52]. The error box area was about 274 sq. arcmin and contained a nearby Sculptor galaxy (3.5 Mpc), indicating possible association of the burst with the magnetar giant flare. The hypothesis was also supported by several features (e.g., by atypically hard the energy spectrum), which were found to be consistent with the soft gamma repeater giant flare hypothesis [8,14,15,38,52,55,56].

The nature of the source is still not clear because of lack of other useful observations (e.g. no optical counterpart was discovered, no precise localization except IPN triangulation was performed).

4.2   Analyzing the $E_{p,i} - E_{iso}$ Correlation

Using spectral data for this burst, obtained by Konus-Wind [13], we build the trajectory for GRB 200415A on the $E_{p,i} - E_{iso}$ diagram (Fig. 1). The trajectory crosses the upper border of $2\sigma_{cor}$ correlation region for type I bursts at $z = 0.06$ and the corresponding correlation fit at $z = 0.56$. It does not cross the lower correlation region border for type I bursts and the upper correlation region border for type II bursts at any redshift, confirming the possible association of the event with type I bursts.

We also compare the position of the burst assuming the $D_L = 3.5$ Mpc of Sculptor galaxy for GRB 200415A with the ones of the SGR giant flares, and find them comparable.

If GRB 200415A is a real type I gamma-ray burst, its redshift should be $z > 0.06$ with the most probable value of $z = 0.56$. Nevertheless, we can not exclude possible association of the burst with the magnetar giant flare.

4.3   Analyzing the $T_{90,i} - EH$ Diagram

Using the duration value of $T_{90} = 0.2\,\text{s}$ [45] and spectral data for this burst [13], we build the trajectory for GRB 200415A on the $T_{90,i} - EH$ diagram (Fig. 2). The trajectory is placed sufficiently above the separation line $EHD = 2.6$ at any redshift, confirming association of the event with type I bursts. As a consequence, we can not estimate any limits on redshift for the burst using the EHD value alone.

The trajectory crosses the borders of 2σ confidence region of type I bursts cluster at $z = 0.035$ and $z = 8.1$, the most probable value is $z = 0.41$. The position of the event at $D_L = 3.5$ Mpc (probable association with Sculptor galaxy) is very close to the positions of SGR giant flares, confirming the association of GRB 200415A with magnetar giant flare, if its source is really placed in Sculptor galaxy.

4.4   Discussion

The most probable values of redshift, obtained by different methods, are comparable: $z = 0.56$ (the $E_{p,i} - E_{iso}$ correlation) and $z = 0.41$ (the $T_{90,i} - EH$ diagram).

Although the $T_{90,i} - EH$ diagram allows us to estimate the upper limit, $z = 8.1$ at 2σ confidence level, this value is rather useless, because type I bursts are not observed at $z > 3$ due to selection effects and progenitor features [34].

Obtained using the $E_{p,i} - E_{iso}$ correlation lower limit on redshift is 1.5 times higher than one obtained using the $T_{90,i} - EH$ diagram ($z = 0.06$ and $z = 0.035$, correspondingly). So, the $E_{p,i} - E_{iso}$ correlation demonstrates better reliability for redshift estimation for this burst.

The possibility of distinguishing type I bursts from SGR giant flares using the $E_{p,i} - E_{iso}$ correlation is considered also in [8,38,56]. While the giant flare of magnetar could reproduce the energetics of the weakest type I bursts (e.g. GRB 170817A), it demonstrates sufficiently harder energy spectrum in terms of $E_{p,i}$, typical for much brighter type I bursts.

5   GRB 200422A

5.1   Observations

The long duration ($T_{90} = 12.3$ s) very bright GRB 200422A was registered at 07:22:16 UT on 22 Apr 2020 by several gamma-ray experiments [18,31,51]. As for early discussed GRB 200415A, the IPN triangulation localization was performed, but gave much weaker result: the error box area was as large as 1 sq. deg [19]. Unfortunately, no x-ray and optical component were discovered, and redshift was not determined.

5.2   Analyzing the $E_{p,i} - E_{iso}$ Correlation

Using spectral data for this burst, obtained by Konus-Wind [51], we build the trajectory for GRB 200422A on the $E_{p,i} - E_{iso}$ diagram (Fig. 1). The trajectory crosses correlation regions of both type I and type II bursts, so we can not classify the burst using just the $E_{p,i} - E_{iso}$ correlation. Nevertheless, the duration value of $T_{90} = 12.3$ s with possible extended emission up to 250 s [31] definitely classifies the burst as the type II one. The trajectory crosses the upper border of $2\sigma_{cor}$ correlation region for type II bursts at $z = 0.09$ and the corresponding correlation fit at $z = 0.70$. Because of the extreme brightness of the burst, we also can estimate the upper limit on redshift as $z = 2.2$, assuming the burst to be not brighter than the brightest known type II bursts, having $E_{iso} \leq 6 \times 10^{54}$ erg.

To conclude, we estimate redshift of the burst to be in the interval $z = (0.09, 2.2)$ with the most probable value of $z = 0.70$.

5.3   Analyzing the $T_{90,i} - EH$ Diagram

Using the duration value of $T_{90} = 12.3\,\mathrm{s}$ [31] and spectral data for this burst [51], we build the trajectory for GRB 200422A on the $T_{90,i} - EH$ diagram (Fig. 2). The duration of the burst definitely characterizes it as the type II one, the intersection point of the trajectory with the separation line $EHD = 2.6$ gives the lower limit of redshift as $z = 0.016$.

The most probable value of redshift is $z = 0.52$. The trajectory crosses the higher border of 2σ confidence region for type II bursts cluster at $z = 0.052$, the intersection with the lower border is not reached within $z \leq 10$ interval, meaning no confident higher limit for redshift for this burst.

5.4   Discussion

An advantage of the $T_{90,i} - EH$ diagram is in a blind classification of the burst. Using just the $E_{p,i}$, E_{iso} parameters gives no possibility to classify this burst, because its trajectory on the $E_{p,i} - E_{iso}$ plane crosses both correlation regions for type I and type II bursts (see Fig. 1).

The $E_{p,i} - E_{iso}$ correlation gives better results in estimating redshift for GRB 200422A. The most probable value is $z = 0.70$ within confidence region of $z = (0.09, 2.2)$. The $T_{90,i} - EH$ diagram gives a comparable value of the most probable redshift, $z = 0.52$ without upper limit on redshift. The low limit on redshift, $z = 0.052$, is twice weaker than one estimated using the $E_{p,i} - E_{iso}$ correlation.

6   Conclusions

We analyze the possibility of using the $E_{p,i} - E_{iso}$ correlation and its derivative, the $T_{90,i} - EH$ diagram to classify gamma-ray bursts and to estimate their redshift.

The $E_{p,i} - E_{iso}$ correlation is shown to be more reliable for estimating the redshift, but the $T_{90,i} - EH$ diagram – to be more reliable in a blind classification.

The possibility of distinguishing type I bursts from SGR giant flares using the $E_{p,i} - E_{iso}$ correlation and the $T_{90,i} - EH$ diagram is also considered.

We classify GRB 200415A as the type I burst with the most probable redshift of $z = 0.56$, with the lower limit of $z = 0.06$ at 2σ confidence level. We confirm the connection of the event with a magnetar giant flare, if the association of the burst with Sculptor galaxy is correct.

We classify GRB 200422A as the type II burst with the most probable redshift of $z = 0.70$, with the lower limit of $z = 0.09$ at 2σ confidence level. The higher limit on redshift, $z = 2.2$ is estimated using the higher limit on energetics of the burst.

Acknowledgments. Authors acknowledge financial support from RSF grant 18-12-00378.

References

1. Abbott, B.P., et al.: GW190425: observation of a compact binary coalescence with total mass ∼3.4 M_o. ApJ **892**(1), L3 (2020). https://doi.org/10.3847/2041-8213/ab75f5
2. Abbott, B.P., et al.: Gravitational waves and gamma-rays from a binary neutron star merger: GW170817 and GRB 170817A. ApJ **848**, L13 (2017). https://doi.org/10.3847/2041-8213/aa920c
3. Abbott, B.P., et al.: Multi-messenger observations of a binary neutron star merger. ApJ **848**, L12 (2017). https://doi.org/10.3847/2041-8213/aa91c9
4. Amati, L., et al.: Intrinsic spectra and energetics of BeppoSAX gamma-ray bursts with known redshifts. A&A **390**, 81–89 (2002). https://doi.org/10.1051/0004-6361:20020722
5. Barkov, M.V., Pozanenko, A.S.: Model of the extended emission of short gamma-ray bursts. MNRAS **417**, 2161–2165 (2011). https://doi.org/10.1111/j.1365-2966.2011.19398.x
6. Blinnikov, S.I., Novikov, I.D., Perevodchikova, T.V., Polnarev, A.G.: Exploding neutron stars in close binaries. Sov. Astron. Lett. **10**, 177–179 (1984)
7. Cano, Z., Wang, S.Q., Dai, Z.G., Wu, X.F.: The observer's guide to the gamma-ray burst supernova connection. Adv. Astron. **2017**, 8929054 (2017). https://doi.org/10.1155/2017/8929054
8. Chand, V., et al.: Magnetar giant flare originated GRB 200415A: transient GeV emission, ten-year Fermi-LAT upper limits and implications. arXiv e-prints arXiv:2008.10822, August 2020
9. Connaughton, V.: BATSE observations of gamma-ray burst tails. ApJ **567**, 1028–1036 (2002). https://doi.org/10.1086/338695
10. Dezalay, J.P., Barat, C., Talon, R., Syunyaev, R., Terekhov, O., Kuznetsov, A.: Short cosmic events: a subset of classical GRBs? In: Paciesas, W.S., Fishman, G.J. (eds.) American Institute of Physics Conference Series. American Institute of Physics Conference Series, vol. 265, p. 304, January 1992
11. Eichler, D., Levinson, A.: An interpretation of the $h\nu_{peak}$-E_{iso} correlation for gamma-ray bursts. ApJ **614**, L13–L16 (2004). https://doi.org/10.1086/425310
12. Fermi GBM Team: GRB 200415A: fermi GBM final real-time localization. GRB Coordinates Netw. **27579**, 1 (2020)
13. Frederiks, D., et al.: Konus-wind team: konus-wind observation of GRB 200415A (a magnetar giant flare in sculptor galaxy?). GRB Coordinates Netw. **27596**, 1 (2020)
14. Frederiks, D.D., et al.: Giant flare in SGR 1806–20 and its Compton reflection from the Moon. Astron. Lett. **33**(1), 1–18 (2007). https://doi.org/10.1134/S106377370701001X
15. Frederiks, D.D., Palshin, V.D., Aptekar, R.L., Golenetskii, S.V., Cline, T.L., Mazets, E.P.: On the possibility of identifying the short hard burst GRB 051103 with a giant flare from a soft gamma repeater in the M81 group of galaxies. Astron. Lett. **33**(1), 19–24 (2007). https://doi.org/10.1134/S1063773707010021

16. Galama, T.J., et al.: An unusual supernova in the error box of the γ-ray burst of 25 April 1998. Nature **395**, 670–672 (1998). https://doi.org/10.1038/27150
17. Gehrels, N., et al.: A new γ-ray burst classification scheme from GRB060614. Nature **444**, 1044–1046 (2006). https://doi.org/10.1038/nature05376
18. Gupta, S., et al.: AstroSat CZTI collaboration: GRB 200422A: AstroSat CZTI detection. GRB Coordinates Netw. **27630**, 1 (2020)
19. Hurley, K., et al.: IPN triangulation of GRB 200422A (long/very bright). GRB Coordinates Netw. **27626**, 1 (2020)
20. Koshut, T.M., et al.: Systematic effects on duration measurements of gamma-ray bursts. ApJ **463**, 570 (1996). https://doi.org/10.1086/177272
21. Kouveliotou, C., et al.: Identification of two classes of gamma-ray bursts. ApJ **413**, L101–L104 (1993). https://doi.org/10.1086/186969
22. Kouveliotou, C., et al.: Discovery of a magnetar associated with the soft gamma repeater SGR 1900+14. ApJ **510**(2), L115–L118 (1999). https://doi.org/10.1086/311813
23. Levinson, A., Eichler, D.: The difference between the Amati and Ghirlanda relations. ApJ **629**, L13–L16 (2005). https://doi.org/10.1086/444356
24. Lü, H.J., Liang, E.W., Zhang, B.B., Zhang, B.: A new classification method for gamma-ray bursts. ApJ **725**(2), 1965–1970 (2010). https://doi.org/10.1088/0004-637X/725/2/1965
25. Marisaldi, M., Mezentsev, A., Østgaard, N., Reglero, V., Neubert, T.: ASIM team: GRB 200415A: ASIM observation. GRB Coordinates Netw. **27622**, 1 (2020)
26. Mazets, E.P., et al.: A giant flare from a soft gamma repeater in the andromeda galaxy (M31). ApJ **680**(1), 545–549 (2008). https://doi.org/10.1086/587955
27. Mazets, E.P., et al.: Catalog of cosmic gamma-ray bursts from the KONUS experiment data. I. Ap&SS **80**, 3–83 (1981). https://doi.org/10.1007/BF00649140
28. Mészáros, P.: Gamma-ray bursts. Rep. Prog. Phys. **69**, 2259–2321 (2006). https://doi.org/10.1088/0034-4885/69/8/R01
29. Meszaros, P., Rees, M.J.: Tidal heating and mass loss in neutron star binaries - implications for gamma-ray burst models. ApJ **397**, 570–575 (1992). https://doi.org/10.1086/171813
30. Metzger, B.D., Quataert, E., Thompson, T.A.: Short-duration gamma-ray bursts with extended emission from protomagnetar spin-down. MNRAS **385**, 1455–1460 (2008). https://doi.org/10.1111/j.1365-2966.2008.12923.x
31. Minaev, P., Chelovekov, I., Pozanenko, A., Grebenev, S.: GRB IKI FuN: GRB 200422A: INTEGRAL observations. GRB Coordinates Netw. **27694**, 1 (2020)
32. Minaev, P.Y., Grebenev, S.A., Pozanenko, A.S., Molkov, S.V., Frederiks, D.D., Golenetskii, S.V.: GRB 070912 - a gamma-ray burst recorded from the direction to the galactic center. Astron. Lett. **38**, 613–628 (2012). https://doi.org/10.1134/S1063773712100064
33. Minaev, P.Y., Pozanenko, A.S.: Precursors of short gamma-ray bursts in the SPI-ACS/INTEGRAL experiment. Astron. Lett. **43**, 1–20 (2017). https://doi.org/10.1134/S1063773717010017
34. Minaev, P.Y., Pozanenko, A.S.: The $E_{p,i}$-E_{iso} correlation: type I gamma-ray bursts and the new classification method. MNRAS **492**(2), 1919–1936 (2020). https://doi.org/10.1093/mnras/stz3611
35. Minaev, P.Y., Pozanenko, A.S., Loznikov, V.M.: Extended emission from short gamma-ray bursts detected with SPI-ACS/INTEGRAL. Astron. Lett. **36**, 707–720 (2010). https://doi.org/10.1134/S1063773710100026

36. Minaev, P.Y., Pozanenko, A.S., Loznikov, V.M.: Short gamma-ray bursts in the SPI-ACS INTEGRAL experiment. Astrophys. Bull. **65**, 326–333 (2010). https://doi.org/10.1134/S1990341310040024

37. Minaev, P.Y., Pozanenko, A.S., Molkov, S.V., Grebenev, S.A.: Catalog of short gamma-ray transients detected in the SPI/INTEGRAL experiment. Astron. Lett. **40**, 235–267 (2014). https://doi.org/10.1134/S106377371405003X

38. Minaev, P., Pozanenko, A.: GRB 200415A: magnetar giant flare or short gamma-ray burst? Astron. Lett. **46**, 573–585 (2020). https://doi.org/10.1134/S1063773720090042

39. Norris, J.P., Bonnell, J.T., Kazanas, D., Scargle, J.D., Hakkila, J., Giblin, T.W.: Long-lag, wide-pulse gamma-ray bursts. ApJ **627**, 324–345 (2005). https://doi.org/10.1086/430294

40. Norris, J.P., Gehrels, N., Scargle, J.D.: Threshold for extended emission in short gamma-ray bursts. ApJ **717**, 411–419 (2010). https://doi.org/10.1088/0004-637X/717/1/411

41. Omodei, N., et al.: Fermi-LAT collaboration: GRB 200415A: Fermi-LAT detection. GRB Coordinates Netw. **27586**, 1 (2020)

42. Paczynski, B.: Gamma-ray bursters at cosmological distances. ApJ **308**, L43–L46 (1986). https://doi.org/10.1086/184740

43. Paczyński, B.: Are gamma-ray bursts in star-forming regions? ApJ **494**, L45–L48 (1998). https://doi.org/10.1086/311148

44. Ade, P.A.R., et al.: Planck 2013 results. XVI. Cosmological parameters. A&A **571**, A16 (2014). https://doi.org/10.1051/0004-6361/201321591. Planck Collaboration

45. Pozanenko, A., Minaev, P., Chelovekov, I., Grebenev, S.: GRB 200415A (possible magnetar Giant Flare in Sculptor Galaxy): INTEGRAL observations. GRB Coordinates Netw. **27627**, 1 (2020)

46. Pozanenko, A.S., et al.: GRB 170817A associated with GW170817: multi-frequency observations and modeling of prompt gamma-ray emission. ApJ **852**, L30 (2018). https://doi.org/10.3847/2041-8213/aaa2f6

47. Pozanenko, A.S., Minaev, P.Y., Grebenev, S.A., Chelovekov, I.V.: Observation of the second LIGO/Virgo event connected with a binary neutron star merger S190425z in the gamma-ray range. Astron. Lett. **45**(11), 710–727 (2020). https://doi.org/10.1134/S1063773719110057

48. Qin, Y.P., Chen, Z.F.: Statistical classification of gamma-ray bursts based on the Amati relation. MNRAS **430**, 163–173 (2013). https://doi.org/10.1093/mnras/sts547

49. Rosswog, S.: Fallback accretion in the aftermath of a compact binary merger. MNRAS **376**(1), L48–L51 (2007). https://doi.org/10.1111/j.1745-3933.2007.00284.x

50. Svinkin, D., et al.: Improved IPN error box for GRB 200415A (consistent with the Sculptor Galaxy). GRB Coordinates Netw. **27595**, 1 (2020)

51. Svinkin, D., et al.: Konus-Wind Team: konus-wind observation of GRB 200422A. GRB Coordinates Netw. **27635**, 1 (2020)

52. Svinkin, D., et al.: IPN triangulation of GRB 200415A (possible Magnetar Giant Flare in Sculptor Galaxy?). GRB Coordinates Netw. **27585**, 1 (2020)

53. Volnova, A.A., et al.: Multicolour modelling of SN 2013dx associated with GRB 130702A. MNRAS **467**, 3500–3512 (2017). https://doi.org/10.1093/mnras/stw3297

54. Woosley, S.E.: Gamma-ray bursts from stellar mass accretion disks around black holes. ApJ **405**, 273–277 (1993). https://doi.org/10.1086/172359

55. Yang, J., et al.: GRB 200415A: a short gamma-ray burst from a magnetar giant flare? ApJ **899**(2), 106 (2020). https://doi.org/10.3847/1538-4357/aba745
56. Zhang, H.M., Liu, R.Y., Zhong, S.Q., Wang, X.Y.: Magnetar giant flare origin for GRB 200415A inferred from a new scaling relation. arXiv e-prints arXiv:2008.05097, August 2020
57. Zhang, Z.B., et al.: Spectrum-energy correlations in GRBs: update, reliability, and the long/short dichotomy. PASP **130**(987), 054202 (2018). https://doi.org/10.1088/1538-3873/aaa6af

Databases of Gamma-Ray Bursts' Optical Observations

Alina Volnova[1]($\boxtimes$) (iD), Alexei Pozanenko[1,2,3] (iD), Elena Mazaeva[1] (iD),
Sergei Belkin[1,3] (iD), and Pavel Minaev[1] (iD)

[1] Space Research Institute of the Russian Academy of Sciences (IKI), 84/32 Profsoyuznaya Street, Moscow 117997, Russia
[2] Moscow Institute of Physics and Technology (MIPT), 9 Institutskiy Per., Dolgoprudny 141701, Russia
[3] Higher School of Economics, National Research University, Moscow 101000, Russia

Abstract. Gamma-ray bursts (GRBs) are the most energetic and yet the most mysterious events in the Universe, which are observed in all ranges of electromagnetic spectrum from radio-bandwidth up to very high energy of about TeV energies. Moreover, recently GRBs accompanying gravitational waves produced in binary neutron star merging were discovered. Despite on the GRB is initially registering by space-born gamma-ray omnidirectional detectors or wide field of view telescopes, the most information for GRB investigation obtained by ground based optical observations. Optical observations allow investigating not only the properties of the bursts themselves, but also their ISM surroundings, host galaxies, and phenomena related to GRBs such as core collapsed supernova and kilonova. Bright GRB can give unique information suitable for modeling, but optically bright burst is registered once per 5 years and can be counted on the fingers of one hand. Statistical analysis is still a main instrument of GRB investigation. We give a review of available optical databases of GRBs and present the original database constructed using optical observations of the Space Research Institute GRB Follow-up Network. We provide examples of some statistical properties of the phenomenon, that can be studied using these databases and discuss constructing necessary databases for future investigations.

Keywords: Gamma-ray bursts · Optical counterparts · Optical afterglows · Optical transients · Supernova · Kilonova · Databases

1 Introduction

Gamma-ray bursts (GRBs) are the most violent extragalactic explosions in the Universe, releasing 1048−1054 ergs in gamma rays typically within a few seconds, and up to a few hours in some instances [1]. GRBs may be divided into two classes regarding their nature: GRBs of Type I with duration less than 2 s, harder spectrum, and caused by merging of close binary systems with at least one neutron star; and GRBs of Type II with duration longer than 2 s, softer spectrum, and caused by death of very massive star during core-collapse (e.g. [2, 3]). In the final stages of merger or collapse, a highly

© Springer Nature Switzerland AG 2021
A. Sychev et al. (Eds.): DAMDID/RCDL 2020, CCIS 1427, pp. 148–159, 2021.
https://doi.org/10.1007/978-3-030-81200-3_11

collimated ejecta is released, which has a typical opening angle of a few degrees [4, 5]. An internal dissipation process within the jet is thought to produce prompt gamma-ray emission [6–8], while a longer lived, multiwavelength afterglow is expected to be produced as the jet propagates through the circumstellar medium (of constant density or a stellar-wind-like density [9, 10]).

In 2/3 of cases, GRBs are accompanied with an X-ray afterglow and in about 40% of cases an optical afterglow is discovered (e.g. [11]). The afterglow may last from several hours up to several months (e.g. GRB 030329 [12], GRB 130427A [13]). The light curves of the X-ray and optical afterglows allow astronomers to investigate circumburst medium, properties of the outburst and its progenitor, and determine the distance to the event measuring a distance via cosmological redshift. Nearby long GRBs exhibit the feature of the supernova (SN) of Type Ic in the light curve which confirms the core-collapse nature of these events [14]. The observations of supernova (SNe) allows estimating main physical properties of the progenitor and the ejection (e.g. [15]). Short GRBs show the kilonova [16–18] feature originating from the merging compact binary system of binary neutron star (e.g. [19, 20]). Indeed, one can extrapolate the progenitors of nearby GRBs onto larger distances and it is believed now that the initially found two phenomenological classes are physically confirmed. Both kilonovae and supernovae allows astronomers to investigate the processes of nucleosynthesis and building matter in the Universe.

When the optical afterglow fades out, the host galaxy of the burst may be observed. The study of properties of host galaxies is one of the tools for investigating the environment in which the GRBs form [21]. In the case of absence of the optical afterglow, the observations of the host galaxy may be the only way to estimate the distance to the event [22].

The first optical counterpart of the GRB was registered in 1997 for GRB 970228 [23] and confirmed the extragalactic origin of these phenomena. Since then GRBs were observed by many different instruments in all ranges of the spectrum, and more than 1300 X-ray and about 800 optical afterglows were discovered [11]. These amounts of data collected allow studying statistical properties of the GRBs parameters, which is crucial for modeling of the processes and getting inside the nature of these events.

Optical data are crucial for GRB investigations, because it allows learning physical properties of the circumburst medium and the burst ejecta. Statistical studies of optical properties of a big number of bursts may give constrains on the evolution of the Universe.

The main problem of the optical data is their big volume and their storage may be a big problem of resources, as much as the problem of data access. That is why in most cases raw data are unavailable for public and only results of reduction and specific parameters measurements are published, and raw data may be accessed only with specific request to their owner. Raw images may contain several Gigabytes per observational night, and some small observatories do not have enough resources to store all raw data. Nevertheless, raw data always contain much more information than is extracted, that is why every database containing raw data is extremely valuable.

In this paper, we discuss databases of raw optical observations of the GRBs and databases containing results of statistical studies of their properties, obtained mostly with optical data. We present the unique database of our Space Research Institute Gamma-Ray Burst Follow-up Network (IKI GRB-FuN) starting from 2001 and containing optical data for more than 500 GRBs.

2   Databases in Optic

2.1   Why Databases Are Necessary?

Statistical analysis along with investigations of very bright GRB are the main tools of GRB investigation. Bright GRB can support us by a dense photometric multicolor light curve, spectroscopy, and polarimetry. Modelling of the light curves and their fitting with empirical equations allow estimating some main characteristics, like time of the break, light curve slopes before and after the break, and investigations of spectra provides photon and spectral indexes. However, bright bursts in optic can be counted on the fingers of one hand. Moreover, the simultaneous detection of prompt phase in optic is still very rare and sparse. Statistical analysis of many combined light curves until now is the main instrument of investigation of a «poor man», and the distributions of the main light curve parameters allow estimating the GRB physical properties according to suggested models. To have a robust data for statistical investigation one need to have a verified data collections/databases.

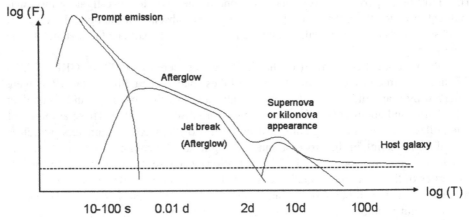

Fig. 1. A schematic light curve of the optical counterpart of a GRB. Main common phases are shown. In some cases, the afterglow phase may be non-monotonic, i.e. contain significant deviations from a power law decay (e.g. [24, 25]).

General optical light curves of GRB consist of several common elements originated from the physics of the phenomenon. Figure 1 shows a scheme of a typical light curve of the GRB optical counterpart. Every marked phase of the light curve originates from different physical processes, and statistical study of every phase allows determining main physical properties characterizing the phase. Statistical dependences inherent in the prompt emission phase are linked to the processes of the burst itself, so it helps to study the burst energetics (e.g. [26, 27]). Many investigations to search for and to study the statistical properties of afterglows and jet breaks have been carried out based the optical data (e.g., [28, 29]). SNe related to the GRBs are also collected to study their statistical properties [30–32]. Statistical properties of short GRBs are collected in a few papers [2, 33, 34]. Till now no dependences are found between prompt gamma-ray emission parameters and optical properties of GRB. Therefore, special interest suggests databases comprising both gamma-ray and optic data (e.g. [35]).

2.2 Available Optical Databases

Below we provide a brief description of the databases and works consisting of the collection of GRB observations in the optical and infrared. All the mentioned databases are quite inhomogeneous in comparison with each other. As comparison criteria, one may consider the total number of GRBs in the database, how many instruments obtained the data (one or many), and does it contain raw data or not. We caution that we have provided mostly usable databases and do not cover all available databases.

GCN: The Gamma-ray Coordinates Network. This is not only a database, it is also an automatic alert service. The network distributes: (i) the Notices - locations of all GRBs and other high energy Transients detected by spacecrafts (most in real-time while the burst is still bursting and others are that delayed due to telemetry down-link delays); and (ii) the Circulars - reports of follow-up observations made by ground-based and space-borne optical, radio, X-ray, TeV, observers. All GCN circulars are in original form, including report about optical observation, so in some cases, it may contain raw data, but mostly it contains results of quick data processing. The database is arranged by GCN number and unique name of the event. The database is continuously updating [36].

GRBGEN – J. Greiner' Database. The base includes only well localized GRBs in gamma-rays (usually better than few arcminutes). The collection is arranged by unique GRB name and includes GCN, ATEL, IAU circulars in original form issued about the GRB, including optical data, and reference to original papers about particular GRBs, if available. The database contains > 2000 GRBs since 1997, when first GRB was localized in X-ray and then in optic [11]. This database is similar to the GCN, but it excludes all non-GRB events, contains a pivot table of all main GRB parameters (coordinates, discovery of different afterglows, redshift, and the discovery gamma-ray observatory), and includes all published results besides quick circulars.

SWIFTGRB - Swift Gamma Ray Bursts Catalog. A rich database of broadband afterglow light curves is accumulating after the Swift satellite was launched in the end of 2004. Since then this observatory became the main facility of discovering and quick observing GRBs. The database contains data collected during the period from the launch of the mission to the end of 2012 from the gamma Burst Alert Telescope, the X-ray telescope (XRT), and the Ultra-Violet Optical Telescope (UVOT), including raw and preprocessed images, localizations, and light curves [37–39]. The main feature is that the database contains data obtained by the only one instrument, so it may miss some events that are not detected by Swift.

Kann's Golden Sample. The two papers written by D.A. Kann [2, 40] contain optical light curves of 79 long and 39 short GRBs, respectively. The data are collected from UVOT/Swift observatory and several ground-based observatories. Statistical properties of long GRBs are compared to those of the short ones. The collection of light curves can be used as templates of GRB afterglows.

MASTER Global Network Early Optical Observations. The catalogue contains the results of early observations for 130 error-boxes of gamma-ray bursts performed with the Mobile Astronomical System of TElescope-Robots (MASTER [41]) global network of robotic telescopes from Moscow State University in fully automatic mode (2011–2017) [42].

The Robotic Optical Transient Search Experiment (ROTSE-III) Catalogue. ROTSE-III is a multi-telescope experiment designed to observe the optical afterglow of gamma-ray bursts and other transient sources like SNe and kilonovae [43]. The catalogue contains original homogeneous data from ROTSE telescopes for 58 GRBs, which are 18 detections and 40 upper limits [44, 45].

A Comprehensive Statistical Study of Gamma-Ray Bursts. [35]. Here the authors tried to collect all of the data and keep them in a reservoir for future studies. The data were all manually collected from the literature, including almost all of the properties belonging to GRBs, i.e., prompt emission, afterglows, and host galaxies. The paper contains all of the possible data for 6289 GRBs, from GRB 910421 (1991 April 21) to GRB 160509A (2016 May 9). There are 46 parameters in this catalog, including the basic information, the prompt emission, the afterglow, and the host galaxy. Now it is the most comprehensive database of the GRB optical studies. Still no raw data are available.

A Large Catalog of Multiwavelength GRB Afterglows. [46]. In the paper, the authors present a large comprehensive catalog of 70 GRBs with multiwavelength optical transient data on which performed a systematic study to find the temporal evolution of color indices. The authors categorize them into two samples based on how well the color indices are evaluated. The Golden sample includes 25 bursts mostly observed by GROND, and the Silver sample includes 45 bursts observed by other telescopes.

GRB-SNe. There are several work collecting observational data on SNe associated with GRB [30–32]. They totally contain of about 40 light curves data, optical spectra and modeling of physical parameters like time of the light curve maximum, the absolute magnitude in the maximum, mass of the progenitor, burst energetics, mass of the ejected Ni56 and some more.

HST Observations of Swift long GRBs. The catalogue contains Hubble Space Telescope (HST) observations of 105 optical afterglows and host galaxies of long GRBs spanning from redshift of 0.03 to 9.4, which were localized using relative astrometry from ground- and space-based afterglow observations [47]. The authors measure the distribution of LGRB offsets from their host centers and their relation to the underlying host light distribution.

GHostS – GRB Host Studies. [48] contains the list of publications, which presents studies of GRB host galaxies in the period from 1997 to 2015. It collects 432 papers about 245 host galaxies.

2.3 IKI GRB-FuN Observations and Data Collection (Database)

The Space Research Institute Gamma-Ray Burst Follow-up Network (IKI GRB-FuN) started operation in 2001. The main idea of the network is using dedicated time on the existed facilities, i.e. IKI GRB-FuN is overlay network. A core of the network in Space Research Institute (IKI) is automatically sending target-of-opportunity applications to different observatories, planning regular observation of afterglow, searching for SN featuring photometric light curve and in spectrum, searching and observing host galaxies. The most prominent result of the IKI GRB-FuN is the observation of GRB 170817 related to the first detection of gravitational-wave event GW170817 of binary neutron star merging [19, 20]. Nowadays the network comprises of about 25 telescopes with aperture from 0.2 to 2.6 m located in different observatories all over the world; the IKI GRB-FuN is also collaborating with ISON network [49] and other observatories by submitting proposals for large aperture telescopes (Table 1). The distribution of the observatories with the longitude allows observing the optical afterglow of the GRB almost without interruption throughout the day and building detailed light curves. All the received data are collected in the Space Research Institute in Moscow.

The database of IKI GRB-FuN comprise multicolor image observations of different optical transients: Gamma-ray bursts optical counterparts, Tidal Disruption Event (TDE), Supernova (SN), Soft Gamma-Rays repeaters (SGR), unclassified transients and observations of the region localization of Gravitational Wave (GW) events registering by LIGO/Virgo in O2 and O3 runs [50]. Database counts more than 500 GRBs with at least one observation available, and 20% of the objects have light curves with more than 10 photometry data. The observations obtained mostly by telescope/observatories that are given in Table 1 and include observations of GRBs in different phases: search for optical counterpart, a few prompt observations, early and late time afterglow observations, supernovae (13) and candidate in supernovae associated with GRBs (4), kilonovae (3), and host galaxies of GRBs (48). Database contains raw data of series observations of the object, calibration data and calibrated a stack image of time series per each epoch. Besides of raw data database contains preliminary photometry of object or an upper limit at the place of supposed object.

Many observations of GRB results in upper limit only, which ranges from 16 m up to 24 m in R-filter. The most of observations performed in R-filter, which is accepted de-facto for observations of GRB afterglow. Using the same filter permits to construct uniform light curve if observations produced by different telescopes/observatories. It is essential to use the same calibration stars for differential photometry for all observations of an object in different telescopes. Calibration stars in the field of view is determined by automatic procedure, which we developed for uniform photometric calibration [51]. The calibration stars are also included in our database. The access to the database may be arranged via FTP protocol, and a specific username and password should be requested by the e-mail to the database PI Alexei Pozanenko (apozanen@iki.rssi.ru).

Table 1. The list of telescopes/observatories used in observations by IKI GRB-FuN database.

Observatory	Telescope name	Diameter, m	Location
Crimean Astrophysical Observatory	ZTSh	2.6	Crimea, Russia
"_"	AZT-11	1.25	"_"
"_"	AZT-8	0.7	"_"
Crimean observatory Koshka (INASAN)	Zeiss-1000	1.0	"_"
Sayan Solar Observatory	AZT-33 IK	1.5	Mondy, Russia
Tein-Shan Astronomical Observatory	Zeiss-1000 East	1.0	Kazakhstan
"_"	Zeiss-1000 West	1.0	"_"
Assy-Turgen Observatory	AZT-20	1.5	"_"
Maidanak Astrophysical Observatory	AZT-22	1.5	Uzbekistan
Burakan Astrophysical Observatory	ZTA	2.6	Armenia
Abastumani Astrophysical Observatory	AS-32	0.7	Georgia
Hureltogoot Observatory	ORI-40, VT-78a	0.4	Mongolia
ISON-Kislovodsk	K-800	0.8	Kislovodsk, Russia
Special Astrophysical Observatory	Zeiss-1000	1.0	N.Arkhyz, Russia
Mt. Terskol Observatory (INASAN)	Zeiss-600	0.6	Terskol, Russia
"_"	Zeiss-2000	2.0	"_"
CHILESCOPE	RC-1000	1.0	Ovalle, Chile
"_"	Newtonian-50	0.5	"_"
South-African Astronomical Observatory	SALT	10.0	Sutherland, South Africa
"_"	40 in.	1.0	"_"
Aryabhatta Research Institute of Observational Sciences	DOT	3.6	Nainital, India
Siding Spring Observatory	PROMPT-Australia	1.0	Australia
Roque de los Muchachos Observatory	GTC	10	Canary Islands
Gemini Observatory	Gemini North	8.1	Hawaii

3   Discussion

In the section, we discuss most important requirements for databases of GRB observations in optic.

3.1   Criteria and Databases Following These Criteria

In discussion of a database of GRB in optic, we would like to follow criterion suggested in [52, 53]. The most sensitive issues are open access, first-hand presentation of data by the experimenter (i.e., original), and uniformity. The two first statements are self explained. The uniformity of the data relies heavily on the use of the same instrument to obtain experimental data. In addition, completeness is important. The notion of completeness of sampling implies the uniformity of data to a certain limit of sensitivity, and the estimating of possible selective effects that impede this uniformity.

From database listed above as uniform and original are Swift, HST, ROTSE-III, MASTER and SN Ic database obtained by ZTF [31]. It is obviously future transient database based on LSST [54], observations give us possibility to blind search for GRB orphans, i.e. gamma-ray burst source not detected in gamma-rays but observed in optic. This is still mysterious but supposed most populated GRB sources because of observed off-axis from the narrow jet which supply energy for prompt emission and early phase of afterglow. It is no doubt that LSST database for optical transients give a kick for targeted search for gamma-ray counterparts for transients found by LSST. Probably, the first GRB orphan already registered by ZTF is ZTF19abvizsw/AT2019pim [55].

3.2   Covering by Observations All Phases of GRB Emission

There are different phases of GRB emission. Prompt phase should be recorded with fine time sampling comparable with time scale of elementary structures (pulses) of gamma-ray observed in gamma-rays. These are the most desirable, but also the most technically sophisticated observations. There are no such observations yet.

The phase of SN associated with GRB is still not well represented. It has natural reasons. SN can be effectively detected up to a distance equivalent to a redshift of z~0.5. However the number of GRBs up to the distance is small, about 2 per year (cf. 100 GRBs per year registered in gamma-rays). From another hand to extract correctly SN light curve one need to cover almost all phases of GRB emission (see Fig. 1) and then to subdivide afterglow, host galaxy and finally obtain the SN light curve. Not in every case we can cover all phases with sufficient sensitivity. For example due to bright Moon at the time of SN appearances prevent precise photometry very often. The multicolor photometry database for afterglow and SN are still not developed. Dense multicolor light curve permit statistical investigation of afterglow light curve (e.g. [2, 40]), in particular to search for inhomogenities in a light curve (e.g. [25]). Investigation of inhomogenities may shed light on the central engine of GRB and properties of circumburst medium.

3.3   Multiwavelength Observations

Despite on the GRB initially is detected by gamma-ray telescopes, emission from a source of GRB registered virtually by in every energy wavelength, from radio- to TeV emission. In a few cases, detected electromagnetic emission spans over 14 orders. It is obvious that synchronous observations in radio, infrared, optic, X-ray and gamma-rays give a lot of information for modeling physical processes of gamma-ray bursts. However till now there is no a comprehensive database comprising all wavelengths. Compiling such a database is a matter of the near future. Of course the database cannot be uniform and complete. However such database is a crucial for comprehensive study of physical processes in GRB, and identifying emission mechanisms at different phases of the GRB. Another problem is a search some connection between properties of prompt emission, afterglow and supernova associating GRB. This is a new area and so far no connections have been found between these three phases, therefore, the creation of multiwavelength databases is extremely necessary.

3.4 Searching for GRB Accompanying Gravitational Wave Events

One of the most interesting problems is observations of GRBs related to gravitational wave events discovered by LIGO/Virgo/KAGRA and searching for kilonova. Short duration GRB is expected after binary neutron star merging (BNS). BNS merging already registered by LIGO/Virgo [56]. Until now there is only one lucky GRB 170817 detected both in gamma-rays and in optic as kilonova after GW170817 [19]. The second one GRB 190425 [57] was registered after second BNS GW190425 detected by LIGO/Virgo [58]. In this case the region of source localization was huge and several optical surveys were fulfilled covering only part of the localization area and galaxies in the expected localization volume. IKI GRB FuN is also participated in the search. No optical counterpart was found in any survey. However, this is not surprising, since none of the surveys covered full localizations area, and the source itself was located at a distance of about four times farther than GW 170817. In addition, the target search for an optical analogue in galaxies may not be exhaustive due to the incompleteness of galaxies in the database up to a distance of 160 Mpc to the source GW190425. In any case, one needs to construct database of each survey, which includes details of target and mosaic observation of particular GW error localization area. The task of finding the optical counterpart in huge areas of localization has proved to be more difficult than finding a needle in a haystack.

The appropriate way to solve all the mentioned problems may be in collecting all GRBs available raw and processed data, as much as the numerical results and constructed light curves, into one big database similar to SIMBAD or NED, and this may be a future issue for the Virtual Observatory development.

References

1. Greiner, J., Mazzali, P.A., Kann, D.A., et al.: A very luminous magnetar-powered supernova associated with an ultra-long γ-ray burst. Nature **523**(7559), 189–192 (2015)
2. Kann, D.A., Klose, S., Zhang, B., et al.: The Afterglows of swift-era gamma-ray bursts. II. Type I GRB versus type II GRB optical afterglows. Astrophys. J. **734**(2), 47 p. (2011). id. 96
3. Kumar, P., Zhang, B.: The physics of gamma-ray bursts & relativistic jets. Phys. Rep. **561**, 1–109 (2015)
4. Racusin, J.L., Liang, E.W., Burrows, D.N., et al.: Jet breaks and energetics of swift gamma-ray burst X-Ray afterglows. Astrophys. J. **698**(1), 43–74 (2009)

5. Zhang, B.-B., van Eerten, H., Burrows, D.N., et al.: An analysis of chandra deep follow-up gamma-ray bursts: implications for off-axis jets. Astrophys. J. **806**(1), 11 p. (2015). id. 15
6. Rees, M.J., Meszaros, P.: Unsteady outflow models for cosmological gamma-ray bursts. Astrophys. J. Lett. **430**, L93–L96 (1994)
7. Kobayashi, S., Piran, T., Sari, R.: Can internal shocks produce the variability in gamma-ray bursts? Astrophys. J. **490**, 92–98 (1997)
8. Hu, Y.-D., Liang, E.-W., Xi, S.-Q., et al.: Internal energy dissipation of gamma-ray bursts observed with swift: precursors, prompt gamma-rays, extended emission, and late X-Ray Flares. Astrophys. J. **789**(2), 13 p. (2014). id. 145
9. Mészáros, P., Rees, M.J.: Optical and long-wavelength afterglow from gamma-ray bursts. Astrophys. J. **476**(1), 232–237 (1997)
10. Sari, R., Piran, T.: Cosmological gamma-ray bursts: internal versus external shocks. Mon. Not. Roy. Astron. Soc. **287**(1), 110–116 (1997)
11. J. Greiner's GRB webpage. http://www.mpe.mpg.de/~jcg/grbgen.html. Accessed 30 Mar 2021
12. Lipkin, Y.M., Ofek, E.O., Gal-Yam, A., et al.: The detailed optical light curve of GRB 030329. Astrophys. J. **606**(1), 381–394 (2004)
13. Perley, D.A., Cenko, S.B., Corsi, A., et al.: The afterglow of GRB 130427A from 1 to 10^{16} GHz. Astrophys. J. **781**(1), 21 p. (2014). id. 37
14. Hjorth, J., Bloom, J.S.: The gamma-ray burst - supernova connection. In: Kouveliotou, C., Wijers, R.A.M.J., Woosley, S. (eds.) Chapter 9 in "Gamma-Ray Bursts", Cambridge Astrophysics Series, vol. 51, pp. 169–190. Cambridge University Press, Cambridge (2012)
15. Volnova, A.A., Pruzhinskaya, M.V., Pozanenko, A.S., et al.: Multicolour modelling of SN 2013dx associated with GRB 130702A. Mon. Not. Roy. Astron. Soc. **467**(3), 3500–3512 (2017)
16. Li, L.-X., Paczynski, B.: Transient events from neutron star mergers. Astrophys. J. Lett. **507**(1), L59–L62 (1998)
17. Metzger, B.D., Martínez-Pinedo, G., Darbha, S., et al.: Electromagnetic counterparts of compact object mergers powered by the radioactive decay of r-process nuclei. Mon. Not. R. Astron. Soc. **406**(4), 2650–2662 (2010)
18. Tanvir, N.R., Levan, A.J., Fruchter, A.S., et al.: A 'kilonova' associated with the short-duration γ-ray burst GRB 130603B. Nature **500**(7464), 547–549 (2013)
19. Abbott, B.P., Abbott, R., Abbott T.D., et al.: Multi-messenger observations of a binary neutron star merger. Astrophys. J. Lett. **848**(2), 59 p. (2017). id. L12
20. Pozanenko, A.S., Barkov, M.V., Minaev, P.Yu., et al.: GRB 170817A associated with GW170817: multi-frequency observations and modeling of prompt gamma-ray emission. Astrophys. J. Lett. **852**(2), 18 p. (2018). id. L30
21. Palla, M., Matteucci, F., Calura, F., Longo, F.: Galactic archaeology at high redshift: inferring the nature of GRB host galaxies from abundances. Astrophys. J. **889**(1), 17 p. (2020). id. 4
22. Volnova, A.A., Pozanenko, A.S. Gorosabel, J., et al.: GRB 051008: a long, spectrally hard dust-obscured GRB in a Lyman-break galaxy at z ≈ 2.8. Mon. Not. Roy. Astron. Soc. **442**(3), 2586–2599 (2014)
23. Costa, E., Frontera, F., Heise, J., et al.: Discovery of an X-ray afterglow associated with the γ-ray burst of 28 February 1997. Nature **387**(6635), 783–785 (1997)
24. Yi, S.X, Yu, H., Wang, F. Y., Dai, Z.-G.: Statistical distributions of optical flares from gamma-ray bursts. Astrophys. J. **844**(1), 8 p. (2017). id. 79
25. Mazaeva, E.; Pozanenko, A.; Minaev, P.: Inhomogeneities in the light curves of gamma-ray bursts afterglow. Int. J. Mod. Phys. **D 27**(10) (2018). id. 1844012
26. Amati, L., Guidorzi, C., Frontera, F., et al.: Measuring the cosmological parameters with the E_p,i-$_{Eis}$o correlation of gamma-ray bursts. Mon. Not. Roy. Astron. Soc. **391**(2), 577–584 (2008)

27. Ghirlanda, G., Nava, L., Ghisellini, G., et al.: Gamma-ray bursts in the co-moving frame. Mon. Not. Roy. Astron. Soc. **420**(1), 483–494 (2012)
28. Fong, W., Berger, E., Metzger, B. D., et al.: Short GRB 130603B: discovery of a jet break in the optical and radio afterglows, and a mysterious late-time X-Ray excess. Astrophys. J. **780**(2), 9 p. (2014). id. 118
29. Wang, X.-G., Zhang, B., Liang, E.-W., et al.: Gamma-ray burst jet breaks revisited. Astrophys. J. **859**(2), 22 p. (2018). id. 160
30. Cano, Z.: The observer's guide to the gamma-ray burst-supernova connection. In: Eighth Huntsville Gamma-Ray Burst Symposium, held 24–28 October 2016 in Huntsville, Alabama. LPI Contribution No. 1962, id. 4116 (2016)
31. Modjaz, M., Liu, Y. Q., Bianco, F.B., Graur, O.: The spectral SN-GRB connection: systematic spectral comparisons between type Ic supernovae and broad-lined type Ic supernovae with and without gamma-ray bursts. Astrophys. J. **832**(2), 23 p. (2016). id. 108
32. Lü, H.-J., Lan, L., Zhang, B., et al.: Gamma-ray burst/supernova associations: energy partition and the case of a magnetar central engine. Astrophys. J. **862**(2), 13 p. (2018). id. 130
33. Pandey, S.B., Hu, Y., Castro-Tirado, A.J., et al.: A multiwavelength analysis of a collection of short-duration GRBs observed between 2012 and 2015. Mon. Not. Roy. Astron. Soc. **485**(4), 5294–5318 (2019)
34. Minaev, P.Y., Pozanenko, A.S.: The $E_p, I_{Eis}o$ correlation: type I gamma-ray bursts and the new classification method. Mon. Not. Roy. Astron. Soc. **492**(2), 1919–1936 (2020)
35. Wang, F., Zou, Y.-C., Liu, F., et al.: A comprehensive statistical study of gamma-ray bursts. Astrophys. J. **893**(1), 90 p. (2020). id. 77
36. GCN Circulars. https://gcn.gsfc.nasa.gov/gcn3_archive.html. Accessed 30 Mar 2021
37. SWIFTGRB - Swift Gamma Ray Bursts Catalog. https://heasarc.gsfc.nasa.gov/W3Browse/swift/swiftgrb.html. Accessed 30 Mar 2021
38. Band, D., Matteson, J., Ford, L., et al.: BATSE observations of gamma-ray burst spectra. I. Spectral diversity. Astrophys. J. **413**, 281 (1993)
39. Donato, D., Angelini, L., Padgett, C.A., et al.: The HEASARC swift gamma-ray burst archive: the pipeline and the catalog. Astrophys. J. Suppl. **203**(1), 17 p. (2012). id. 2
40. Kann, D.A., Klose, S., Zhang, B., et al.: The afterglows of swift-era gamma-ray bursts. I. Comparing pre-swift and swift-era long/soft (Type II) GRB optical afterglows. Astrophys. J. **720**(2), 1513–1558 (2010)
41. Lipunov, V., Kornilov, V., Gorbovskoy, E., et al.: Master robotic net. Adv. Astron. (2010). id. 349171
42. Ershova, O.A., et al.: Early optical observations of gamma-ray bursts compared with their gamma- and X-Ray characteristics using a MASTER global network of robotic telescopes from Lomonosov moscow state university. Astron. Rep. **64**(2), 126–158 (2020). https://doi.org/10.1134/S1063772920020018
43. Akerlof, C.W., Ashley, M.C.B., Casperson, D.E., et al.: The ROTSE-III robotic telescope system. Publ. Astron. Soc. Pac. **115**(803), 132–140 (2003)
44. Rykoff, E.S., Aharonian, F., Akerlof, C.W., et al.: Looking into the fireball: ROTSE-III and swift observations of early gamma-ray burst afterglows. Astrophys. J. **702**(1), 489–505 (2009)
45. Cui, X.H., Wu X.F., Wei J.J., et al.: The optical luminosity function of gamma-ray bursts deduced from ROTSE-III observations. Astrophys. J. **795**(2), 6 p. (2014). id. 103
46. Li, L., Wang, Y., Shao, L., et al.: A large catalog of multiwavelength GRB afterglows. I. color evolution and its physical implication. Astrophys. J. Suppl. Ser. **234**(2), 29 p. (2018). id. 26
47. Blanchard, P.K., Berger, E., Fong, W.: The offset and host light distributions of long gamma-ray bursts: a new view from HST observations of swift bursts. Astrophys. J. **817**(2), 30 p. (2016). id. 144
48. GHostS – GRB Host Studies. http://www.grbhosts.org/Default.aspx

49. Pozanenko, A., Mazaeva, E., Volnova, A., et al.: GRB afterglow observations by international scientific optical network (ISON). In: Eighth Huntsville Gamma-Ray Burst Symposium, held 24–28 October, 2016 in Huntsville, Alabama. LPI Contribution No. 1962, id.4074 (2016)
50. Mazaeva, E., et al.: Searching for optical counterparts of LIGO/Virgo events in O2 run. In: Elizarov, A., Novikov, B., Stupnikov, S. (eds.) DAMDID/RCDL 2019. CCIS, vol. 1223, pp. 124–143. Springer, Cham (2020). https://doi.org/10.1007/978-3-030-51913-1_9
51. Skvortsov, N.A., et al.: Conceptual approach to astronomical problems. Astrophys. Bull. **71**(1), 114–124 (2016). https://doi.org/10.1134/S1990341316010120
52. Kalinichenko, L.A., Volnova, A.A., Gordov, E.P., et al.: Data access challenges for data intensive research in Russia. Inform. Appl. **10**(1), 2–22 (2016)
53. Kalinichenko, L., Fazliev, A., Gordov, E.P., et al.: New data access challenges for data intensive research in Russia. In: CEUR Workshop Proceedings "Selected Papers of the 17th International Conference on Data Analytics and Management in Data Intensive Domains, DAMDID/RCDL 2015", vol. 1536, pp. 215–237 (2015)
54. LSST Science Collaborations and LSST Project 2009, LSST Science Book, Version 2.0, arXiv:0912.0201
55. Kool, E., Stein, R, Sharma, Y., et al.: LIGO/Virgo S190901ap: Candidates from the Zwicky Transient Facility. GCN Circ. 25616 (2019)
56. Abbott, B.P., Abbott, R., Abbott, T.D. et al.: Gravitational waves and gamma-rays from a binary neutron star merger: GW170817 and GRB 170817A. Astrophys. J. Lett. **848**(2), 27 p. (2017). id. L13
57. Pozanenko, A.S., Minaev, P.Y., Grebenev, S.A., Chelovekov, I.V.: Observation of the second LIGO/Virgo event connected with a binary neutron star merger S190425z in the gamma-ray range. Astron. Lett. **45**(11), 710–727 (2019). https://doi.org/10.1134/S1063773719110057
58. Abbott, B.P., Abbott, R., Abbott, T.D., et al.: GW190425: observation of a compact binary coalescence with total mass ∼3.4 M⊙. Astrophys. J. Lett. **892**(1), 24 p. (2020). id. L3
59. SIMBAD - SIMBAD Astronomical Database - CDS (Strasbourg). http://simbad.u-strasbg.fr/simbad/. Accessed 30 Mar 2021
60. NED - NASA/IPAC Extragalactic Database. https://ned.ipac.caltech.edu. Accessed 30 Mar 2021
61. Virtual Observatory http://ivoa.net. Accessed 30 Mar 2021

Information Extraction from Text

Part of Speech and Gramset Tagging Algorithms for Unknown Words Based on Morphological Dictionaries of the Veps and Karelian Languages

Andrew Krizhanovsky[1,2]($\boxtimes$) (iD), Natalia Krizhanovskaya[1] (iD), and Irina Novak[3] (iD)

[1] Institute of Applied Mathematical Research of the Karelian Research Centre of the Russian Academy of Sciences, Petrozavodsk, Russia
[2] Petrozavodsk State University, Petrozavodsk, Russia
[3] Institute of Linguistics, Literature and History of the Karelian Research Centre of the Russian Academy of Sciences, Petrozavodsk, Russia
http://dictorpus.krc.karelia.ru

Abstract. This research devoted to the low-resource Veps and Karelian languages. Algorithms for assigning part of speech tags to words and grammatical properties to words are presented in the article. These algorithms use our morphological dictionaries, where the lemma, part of speech and a set of grammatical features (gramset) are known for each word form. The algorithms are based on the analogy hypothesis that words with the same suffixes are likely to have the same inflectional models, the same part of speech and gramset. The accuracy of these algorithms were evaluated and compared. 66 thousand Karelian and 313 thousand Vepsian words were used to verify the accuracy of these algorithms. The special functions were designed to assess the quality of results of the developed algorithms. 86.8% of Karelian words and 92.4% of Vepsian words were assigned a correct part of speech by the developed algorithm. 90.7% of Karelian words and 95.3% of Vepsian words were assigned a correct gramset by our algorithm. Morphological and semantic tagging of texts, which are closely related and inseparable in our corpus processes, are described in the paper.

Keywords: Morphological analysis · Low-resource language · Part of speech tagging

1 Introduction

Our work is devoted to low-resource languages: Veps and Karelian. These languages belong to the Finno-Ugric languages of the Uralic language family. Most Uralic languages still lack full-fledged morphological analyzers and large corpora [4].

In order to avoid this trap the researchers of Karelian Research Centre are developing the Open corpus of Veps and Karelian languages (VepKar). Our

© Springer Nature Switzerland AG 2021
A. Sychev et al. (Eds.): DAMDID/RCDL 2020, CCIS 1427, pp. 163–177, 2021.
https://doi.org/10.1007/978-3-030-81200-3_12

corpus contains morphological dictionaries of the Veps language and the three supradialects of the Karelian language: the Karelian Proper, Livvi-Karelian and Ludic Karelian. The developed software (corpus manager)[1] and the database, including dictionaries and texts, have open licenses.

Algorithms for assigning part of speech tags to words and grammatical properties to words, without taking into account a context, using manually built dictionaries, are presented in the article (see Sect. 4).

The proposed technology of evaluation (see the Sect. 5) allows to use all 313 thousand Veps and 66 thousand Karelian words to verify the accuracy of the algorithms (Table 1). Only a third of Karelian words (28%) and two-thirds of Veps words (65%) in the corpus texts are automatically linked to the dictionary entries with all word forms (Table 1). These words were used in the evaluation of the algorithms.

Table 1. Total number of words in the VepKar corpus and dictionary

Language	The total number of tokens in texts, 10^3	N tokens linked to dictionary automatically, 10^3	N tokens linked to lemmas having a complete paradigm, 10^3
Veps	488	400 (82%)	313 (65%)
Karelian proper	245	111 (45%)	69 (28%)

Let us describe several works devoted to the development of morphological analyzers for the Veps and Karelian languages.

- The Giellatekno language research group is mainly engaged in low-resource languages, the project covers about 50 languages [3]. Our project has something in common with the work of Giellatekno in that (1) we work with low-resource languages, (2) we develop software and data with open licenses. A key role in the Giellatekno infrastructure is given to formal approaches (grammar-based approach) in language technologies. They work with morphology rich languages. Finite-state transducers (FST) are used to analyse and generate the word forms [3].
- There is a texts and words processing library for the Uralic languages called UralicNLP [2]. This Python library provides interface to such Giellatekno tools as FST for processing morphology and constraint grammar for syntax. The UralicNLP library lemmatizes words in 30 Finno-Ugric languages and dialects including the Livvi dialect of the Karelian language (*olo* – language code).

---

[1] See https://github.com/componavt/dictorpus.

2 Data Organization and Text Tagging in the VepKar Corpus

Automatic text tagging is an important area of research in corpus linguistics. It is required for our corpus to be a useful resource.

The corpus manager handles the dictionary and the corpus of texts (Fig. 1). The texts are segmented into sentences, then sentences are segmented into words (tokens). The dictionary includes lemmas with related meanings, word forms, and **sets** of **gram**matical features (in short – **gramsets**).

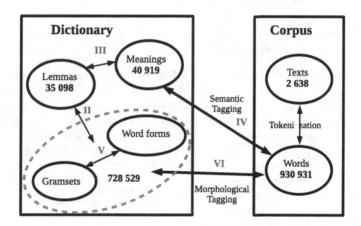

Fig. 1. Data organization and text tagging in the VepKar corpus. Total values (e.g. number of words, texts) are calculated for all project languages.

Text tokens are automatically searched in the dictionary of lemmas and word forms, this is the first stage (I) of the text tagging, it is not presented at Fig. 1.

1. **Semantic tagging.** For the word forms found in the dictionary, the lemmas linked with them are selected (II at Fig. 1), then all the meanings of the lemmas are collected (III) and semantic relationships are established between the tokens and the meanings of the lemmas (marked "not verified") (IV).
   The task of an expert linguist is to check these links and confirm their correctness, either choose the correct link from several possible ones, or manually add a new word form, lemma or meaning.

$$\text{tokens (words)} \xleftrightarrow{\text{I}} \text{word forms} \xleftrightarrow{\text{II}} \text{lemmas} \xleftrightarrow{\text{III}} \text{meanings}$$
$$\underbrace{\qquad\qquad\qquad\qquad\qquad}_{\text{IV not verified}}$$

When the editor clicks on the token in the text, then a drop-down list of lemmas with all the meaning will be shown. The editor selects the correct lemma and the meaning (Fig. 2).

2. **Morphological tagging.** For the word forms found in the dictionary, the gramsets linked with them are selected (V) and morphological links are established (VI) between the tokens and the pairs "word form – gramset" (Fig. 1). The expert's task is to choose the right gramset.

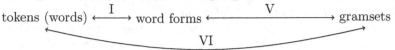

tokens (words) $\xleftrightarrow{\text{I}}$ word forms $\xleftarrow{\hspace{2cm}\text{V}\hspace{2cm}}$ gramsets

VI

Corpus: Biblical texts (translated)

Source: Uz' Zavet, (2006), p. 163-164
Evangelii Lukan mödhe. Iisus matkas jerusalimha 9:51-19:27. 10. От Луки святое благовествование, Глава 10. Библия (Синодальный перевод).

Hüväsüdäimeline samarialaine (Vepsian)

²⁵Eraz käskištonopendai tahtoi kodvda Iisusad. Hän libui i küzui: «Opendai, midä minei tarbiž tehta, miše minä s
elon?» ²⁶Iisus sanu
käskištos? Kut sinä
«Armasta Ižandad
südäimel i kaikel h
melel, i ičeiž lähen
«Oikti sinä sanuid.

opendai (teacher, instructor) +
opeta (1) to teach) +
opeta (2) to convince, to suggest) +
opeta (3) to master, to become proficient) +
opeta (4) to inculcate, to impart, to cultivate) +
opeta (5) ru: наказать, проучить) +

²⁹No mez', ku taht
käsköiden mödhe,
minun lähembaine

³⁰Iisus sanui häne
jerusalimaspäi Jeri...
hänen päle. Hö heitiba hänel sobad-ki pälpäi i

Hüväsüdäimeline samarialaine (Russian)

²⁵ И вот, один законник встал и, искушая Его, сказал: Учитель! чтó мне делать, чтобы наследовать жизнь вечную?
²⁶ Он же сказал ему: в законе что написано? кáк читаешь?
²⁷ Он сказал в ответ: возлюби Господа Бога твоего всем сердцем твоим, и всею душею твоею, и всею крепостию твоею, и всем разумением твоим, и ближнего твоего, как самого себя.
²⁸ Иисус сказал ему: правильно ты отвечал; так поступай, и будешь жить.
²⁹ Но он, желая оправдать себя, сказал Иисусу: а кто мой ближний?
³⁰ На это сказал Иисус: некоторый человек шел из Иерусалима в Иерихон и попался разбойникам, которые сняли с него одежду,

Fig. 2. Vepsian and Russian parallel Bible translation[‡] in the corpus. The editor clicks the word "Opendai" in the text, a menu pops up. This menu contains a list of meanings collected automatically for this token, namely: the meaning of the noun "teacher" ("opendai" in Veps) and five meanings of the Veps verb "opeta". The noun "opendai" and the verb "opeta" have the same wordform "opendai". If the editor selects one of the lemma meanings in the menu (clicks the plus sign), then the token and the correct meaning of the lemma will be connected (IV stage is verified). ([‡] See full text online at VepKar: http://dictorpus.krc.karelia.ru/en/corpus/text/494)

3   Corpus Tagging Peculiarities

In this section, we describe why the word forms with white spaces and analytical forms are not taken into account in the search algorithm described below. Analytical form is the compound form consisting of auxiliary words and the main word.

The ultimate goal of our work is the morphological markup of the text, previously tokenized into words by white spaces and non-alphabetic characters (for example, brackets, punctuation, numbers). Therefore, analytical forms do not have markup in the texts.

Although we store complete paradigms in the dictionary, including analytical forms, such forms do not used in the analysis of the text, because each individual word is analyzed in the text, not a group of words.

For example, we take the Karelian verb "pageta" (leave, run away). In the dictionary not only the negative form of indicative, presence, first-person singular "en pagene" is stored, but also connegative (a word form used in negative clauses) of the indicative, presence "pagene", which is involved in the construction of five of the six forms of indicative, presence. Thus, in the text the word 'en' (auxiliary verb 'ei', indicative, first-person singular) and 'pagene' (verb 'pageta', connegative of indicative, presence) are separately marked.

4   Part of Speech and Gramset Search by Analogy Algorithms

The proposed algorithms operate on data from a morphological dictionary. The algorithms are based on the analogy hypothesis that words with the *same suffixes* are likely to have the same inflectional models and the same sets of grammatical information (part of speech, number, case, tense, etc.). The *suffix* here is a final segment of a string of characters.

Let the hypothesis be true, in that case, if the suffixes of new words coincide with the suffixes of dictionary words, then part of the speech and other grammatical features of the new words will coincide with the dictionary words. It should be noted that the length of the suffixes is unpredictable and can be various for different pairs of words [1, p. 53].

The POSGuess and GramGuess algorithms described below use the concept of "suffix" (Fig. 3), the GramPseudoGuess algorithm uses the concept "pseudo-ending" (Fig. 4).

4.1   The POSGuess Algorithm for Part of Speech Tagging with A Suffix

Given the set of words W, for each word in this set the part of speech is known. The Algorithm 1 finds a part of speech pos_u for a given word u using this set.

In Algorithm 1 we look for in the set W (line 5) the words which have the same suffix u_z as the unknown word u. Firstly, we are searching for the longest substring of u, that starts at index z. The first substring $u_{z=2}$ will start at the second character (line 1 in Algorithm 1), since $u_{z=1} = u$ is the whole string (Fig. 3).

Then we increment the value z decreasing the length of the substring u_z in the loop, while the substring u_z has non-zero length, $z \leq len(u)$. If there are words in W with the same suffix, then we count the number of similar words for each part of the speech and stop the search.

The Fig. 3 shows the idea of the Algorithm 1: for a new word (*kezaman*), we look for a word form in the dictionary (*raman*) with the same suffix (*aman*).

We begin to search in the dictionary for word forms with the suffix $u_{z=2}$. If we have not find any words, then we are looking for $u_{z=3}$ and so on. The longest suffix $u_{z=4} =$ "aman" with $z = 4$ is found.

Then we find all words with the suffix $u_{z=4}$ and count how many of such words are nouns, verbs, adjectives and so on. The result is written to the array *Counter*[]. In Fig. 3 the *noun* "raman" was found, therefore we increment the value of *Counter*[*noun*].

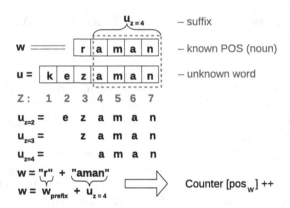

Fig. 3. Veps nouns in the genitive case "kezaman" ("kezama" means "melted ground") and "raman" ("rama" means "frame"). The word u with an unknown part of speech is "kezaman". The word w from the dictionary with the known POS is "raman". They share the common suffix u_z, which is "aman".

Algorithm 1: Part of speech search by a suffix (POSGuess)

Data: P – a set of part of speech (POS),
$W = \{w \mid \exists \text{pos}_w \in P\}$ – a set of words, POS is known for each word,
$u \notin W$ – the word with unknown POS,
$len(u)$ – the length (in characters) of the string u.

Result:

$$u_z : \begin{cases} len(u_z) \xrightarrow[z=2,\ldots,len(u)]{} \max, \text{ // Longest suffix} \\ \exists w \in W : w = w_{\text{prefix}} + u_z \text{ // Concatenation of strings} \end{cases}$$

$$\text{Counter}\left[\text{pos}^k\right] = c^k, \ k = \overline{1,m}, \text{ where :}$$
$$c^k \in \mathbb{N}, \ c^1 \geq c^2 \geq \ldots \geq c^m,$$
$$\exists w_i^k \in W : w_i^k = w_{\text{prefix}_i}^k + u_z \Rightarrow c^k = |\text{pos}_{w_i^k}^k|,$$
$$i = \overline{1, c^k},$$
$$\forall i : \text{pos}_{w_i^k}^k = \text{pos}^k \in P, \quad a \neq b \Leftrightarrow \text{pos}^a \neq \text{pos}^b$$
$$m - \text{the number of different POS of found words } w_i^k$$

---

1 $z = 2$ // The position in the string u
2 $z_{found} = \text{FALSE}$
3 while $z \leq len(u)$ and $\neg z_{found}$ **do**
　　// The suffix of the word u from z-th character
4　　$u_z = \text{substr}(u, z)$
5　　**foreach** $w \in W$ **do**
　　　　// If the word w has the suffix u_z (regular expression)
6　　　　**if** $w =\sim m/u_z\$/$ **then**
7　　　　　Counter $[\text{pos}_w] + +$
8　　　　　$z_{found} = \text{TRUE}$ // Only POS of words with this u_z suffix
　　　　　will be counted. The next "while" loop will break, so the
　　　　　shorter suffix u_{z+1} will be omitted.
9　　　　**end**
10　　**end**
11　　$z = z + 1$
12 end
　　// Sort the array in descending order, according to the value
13 arsort(Counter [ ])

4.2   The GramGuess Algorithm for Gramset Tagging with a Suffix

The GramGuess algorithm is exactly the same as the POSGuess algorithm, except that it is needed to search a subset of gramsets instead of parts of speech.

That is in the set W the gramset is known for each word. The gramset is a set of morphological tags (number, case, tense, etc.).

4.3   The GramPseudoGuess Algorithm for Gramset Tagging with a Pseudo-ending

Let us explain the "pseudo-ending" used in the algorithm GramPseudoGuess.

All word forms of one lemma share a common invariant substring. This substring is a *pseudo-base* of the word (Fig. 4). Here the pseudo-base is placed at the start of a word, it suits for the Veps and Karelian languages. For example, in Fig. 4 the invariant substring "huuk" is the pseudo-base for all word forms of the lemma "huukkua". The Karelian verb "huukkua" means "to call out", "to holler", "to halloo".

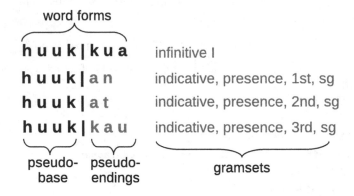

Fig. 4. Wordforms of the Karelian verb "huukkua" (it means "to call out", "to holler", "to halloo"). All word forms have the same pseudo-base and different pseudo-endings for different set of grammatical attributes (gramsets).

Given the set of words W, for each word in this set a gramset and a pseudo-ending are known. The Algorithm 2 finds a gramset g_u for a given word u using this set.

In Algorithm 2 we look for in the set W (line 5) the words which have the same pseudo-ending u_z as the unknown word u. Firstly, we are searching for the longest substring of u, that starts at index z.

Then we increment the value z decreasing the length of the substring u_z in the loop, while the substring u_z has non-zero length, $z \leq len(u)$. If there are words in W with the same pseudo-ending, then we count the number of similar words for each gramset and stop the search.

Algorithm 2: Gramset search by a pseudo-ending (GramPseudoGuess)

Data: G – a set of gramsets,

$W = \{w \mid \exists\, g_w \in G, \exists\, \text{pend}_w : w = w_{\text{prefix}} + \text{pend}_w\}$ – a set of words, where gramset and pseudo-ending (pend) are known for each word,

$u \notin W$ – the word with unknown gramset,

$len(u)$ – the length (in characters) of the string u.

Result:

$$u_z : \begin{cases} len(u_z) \xrightarrow[z=2,\ldots,len(u)]{} \max, & // \texttt{Longest substring} \\ \exists\, w \in W : \text{pend}_w = u_z \end{cases}$$

$$\text{Counter}\left[g^k\right] = c^k, \; k = \overline{1,m}, \; \text{where :}$$

$$c^k \in \mathbb{N}, \; c^1 \geq c^2 \geq \ldots \geq c^m,$$

$$\exists\, w_i^k \in W : \text{pend}_{w_i^k} = u_z \Rightarrow c^k = |g_{w_i^k}^k|,$$

$$i = \overline{1, c^k},$$

$$\forall i : g_{w_i^k}^k = g^k \in G, \quad a \neq b \Leftrightarrow g^a \neq g^b$$

$$m - \text{the number of different gramsets of found words } w_i^k$$

```
1  z = 2 // The position in the string u
2  z_found = FALSE
3  while z ≤ len(u) and ¬z_found do
       // The substring of the word u from z-th character
4      u_z = substr (u, z)
5      foreach w ∈ W do
           // If the word w has the pseudo-ending u_z
6          if pend_w == u_z then
7              Counter [g_w] + +
8              z_found = TRUE // Only gramsets of words with the
                   pseudo-ending u_z will be counted. The next "while" loop
                   will break, so the shorter u_{z+1} will be omitted.
9          end
10     end
11     z = z + 1
12 end
   // Sort the array in descending order, according to the value
13 arsort( Counter [ ] )
```

5   Experiments

5.1   Data Preparation

Lemmas and word forms from our morphological dictionary were gathered to *one set* as a search space of part of speech tagging algorithm. This set contains unique pairs "word – part of speech".

In order to search a gramset, we form the set consisting of (1) lemmas without inflected forms (for example, adverbs, prepositions) and (2) inflected forms (for example, nouns, verbs). This set contains unique pairs "word – gramset". For lemmas without inflected forms the gramset is empty.

We put on constraints for the words in both sets: strings must consist of more than two characters and must not contain whitespace. That is, analytical forms and compound phrases have been excluded from the sets (see Sect. 3).

5.2   Part of Speech Search by a Suffix (POSGuess Algorithm)

For the evaluation of the quality of results of the searching algorithm POSGuess the following function eval(pos^u) was proposed:

$$\text{eval}\left(\text{pos}^u, \text{Counter}\left[\text{pos}^k\right] \rightarrow c^k, \forall k = \overline{1, m}\right) =$$

$$\begin{cases} \text{The array Counter}[] \text{ do not contain the correct pos}^u. \\ 0, \quad \text{pos}^u \neq \text{pos}^k, \forall k = \overline{1, m}, \\[4pt] \text{First several POS in the array can have the same} \\ \text{maximum frequency } c^1, \text{ one of this POS is pos}^u. \\ 1, \quad \text{pos}^u \in \{\left[\text{pos}^1, \ldots, \text{pos}^j\right] : c^1 = c^2 = \ldots = c^j, j \leq m\}, \\[4pt] \frac{c^k}{\sum_{k=1}^m c^k}, \quad \exists k : \text{pos}^k = \text{pos}^u, \ c^k < c^1 \end{cases} \quad (1)$$

This function (1) evaluates the result of the POSGuess algorithm against the correct part of speech pos^u. The POSGuess algorithm counts the number of words similar to the word u separately for each part of speech and stores the result in the Counter array.

The Counter array is sorted in descending order, according to the value. The first element in the array is a part of speech with maximum number of words similar to the unknown word u.

71 091 "word – part of speech" pairs for the Proper Karelian supradialect and 399 260 "word – part of speech" pairs for the Veps language have been used in the experiments to evaluate algorithms.

During the experiments, two Karelian words were found, for which there were no suffix matches in the dictionary. They are the word "cap" (English: snap; Russian: *цап*) and the word "štob" (English: in order to; Russian: *чтобы*). That is, there were no Karelian words with the endings -p and -b. This could be explained by the fact that these two words migrated from Russian to Karelian language.

Table 2 (Veps) and Table 3 (Karelian) show the evaluation of the results of the POSGuess algorithm for each part of speech. The function for evaluating the part of speech assignment (see the formula (1)) takes the following values:

0   4.7% of Vepsian words and 9% of Karelian words were assigned the wrong part of speech. That is, there is no correct part of speech in the result array $Counter[]$ in the POSGuess algorithm. This is the first line in the formula (1).

0.1 – 0.5   2.92% of Vepsian words and 4.23% of Karelian words were assigned the partially correct POS tags. That is, the array $Counter[]$ contains the correct part of speech, but it is not at the beginning of the array. This is the last line in the formula (1).

1   92.38% of Vepsian words and 86.77% of Karelian words were assigned the correct part of speech. The array $Counter[]$ contains the correct part of speech at the beginning of the array.

Table 2. Number of Vepsian words of different parts of speech used in the experiment. The evaluation of results found by POSGuess algorithm by the formula (1) and fraction of results in percent, where the column 0 means the fraction of words with incorrectly found POS, 1 – the fraction of words with correct POS in the top of the list created by the algorithm.

Veps		Fraction of not guessed (column 0), partly guessed (0.1–0.5) and guessed (1) POS, %						
POS	Words	0	0.1	0.2	0.3	0.4	0.5	1
Verb	93 047	2.12	0.52	0.55	0.47	0.36	0.01	**95.97**
Noun	240 513	2.88	0.3	0.67	0.6	0.42	0.24	**94.89**
Adjective	61 845	12.45	1.62	1.44	1.53	1.58	0.51	**80.87**
Pronoun	1244	46.54	8.12	0.56	0.64	0	0	**44.13**
Numeral	1200	44	6.25	2.33	0.67	0.33	0	**46.42**
Adverb	650	64.92	3.08	2.46	1.23	0.46	0	**27.85**

5.3   Gramset Search by a Suffix (GramGuess Algorithm) and by a Pseudo-ending (GramPseudoGuess Algorithm)

73 395 "word – gramset" pairs for the Karelian Proper supradialect and 452 790 "word – gramset" pairs for the Veps language have been used in the experiments to evaluate GramGuess and GramPseudoGuess algorithms.

A list of gramsets was searched for each word. The list was ordered by the number of similar words having the same gramset.

Table 3. Number of Karelian words of different parts of speech used in the experiment.

Karelian		Fraction of not guessed (column 0), partly guessed (0.1–0.5) and guessed (1) POS, %						
POS	Words	0	0.1	0.2	0.3	0.4	0.5	1
Verb	26 033	3.26	0.5	0.74	0.6	0.23	0.01	**94.67**
Noun	36 908	5.47	0.38	1.13	1.08	0.52	0.04	**91.38**
Adjective	6596	35.81	6.66	4.15	4.56	2.73	0.38	**45.71**
Pronoun	610	81.64	2.13	0.66	3.11	2.3	0	**10.16**
Numeral	582	65.81	1.72	1.03	0.17	1.03	0	**30.24**
Adverb	235	68.51	3.4	2.98	2.13	0	0	**22.98**

Table 4. Evaluations of results of gramsets search for Vepsian and Karelian by GramGuess and GramPseudoGuess algorithms.

Evaluation	GramGuess		GramPseudoGuess	
	Veps	Karelian	Veps	Karelian
0	2.53	5.72	7.9	9.23
0.1	0.53	0.83	1.04	1.57
0.2	0.71	1.16	1.24	1.37
0.3	0.64	0.89	2.68	1.36
0.4	0.2	0.56	0.14	0.68
0.5	0.11	0.09	0.83	0.43
1	**95.29**	**90.74**	**86.17**	**85.36**

For the evaluation of the quality of results of the searching algorithms the following function eval(g^u) has been proposed:

$$\text{eval}\left(g^u, \text{Counter}\left[g^k\right] \to c^k, \forall k = \overline{1,m}\right) =$$

$$
\begin{cases}
\text{The array Counter do not contain the correct gramset } g^u. \\
0, \quad g^u \neq g^k, \forall k = \overline{1,m}, \\[4pt]
\text{First several gramsets in the array can have the same} \\
\text{maximum frequency } c^1, \text{ one of these gramsets is } g^u. \\
1, \quad g^u \in \{[g^1,\dots,g^j] : c^1 = c^2 = \dots = c^j, j \leq m\}, \\[4pt]
\frac{c^k}{\sum_{k=1}^m c^k}, \quad \exists k : g^k = g^u, \ c^k < c^1
\end{cases}
\qquad (2)
$$

This function (2) evaluates the results of the GramGuess and GramPseudoGuess algorithms against the correct gramset g^u.

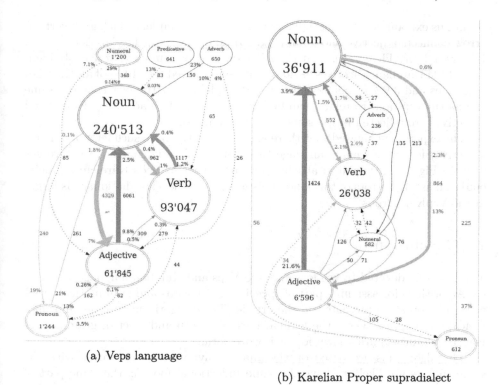

(a) Veps language

(b) Karelian Proper supradialect

Fig. 5. Part-of-speech error transition graph, which reflects the results of the POSGuess algorithm.

The Table 4 shows that the GramGuess algorithm gives the better results than the GramPseudoGuess algorithm, namely:

Karelian   90.7% of Karelian words were assigned a correct gramset by Gram-
           Guess algorithm versus 85.4% by GramPseudoGuess algorithm;

Veps   95.3% of Vepsian words were assigned a correct gramset by GramGuess
           algorithm versus 85.4% by GramPseudoGuess algorithm.

It may be suggested by the fact that suffixes are longer than pseudo-endings. In addition, the GramPseudoGuess algorithm is not suitable for a part of speech without inflectional forms.

6   Morphological Analysis Results

In order to analyze the algorithm errors, the results of the part-of-speech algorithm POSGuess were visualized using the Graphviz program. Part-of-speech error transition graphs were built for Veps language (Fig. 5a) and Karelian Proper supradialect (Fig. 5b).

Let us explain how these graphs were built. For example, a thick grey vertical arrow connects adjective and noun (Fig. 5b), and this arrow has labels of 21.6%, 1424 and 3.9%. This means that the POSGuess algorithm has erroneously identified 1424 Karelian adjectives as nouns. This accounted for 21.6% of all Karelian adjectives and 3.9% of nouns. This can be explained by the fact that the same lemma (in Veps and Karelian) can be both a noun and an adjective. Nouns and adjectives are inflected in the same form (paradigm).

The experiment showed that there are significantly more such lemmas (noun-adjective) for the Karelian language than for the Veps language (21.6% versus 9.8% in Fig. 5). Although in absolute numbers Veps exceeds Karelian, namely: 6061 versus 1424 errors of this kind. This is because the Veps dictionary is larger in the VepKar corpus.

7   Conclusion

This research devoted to the low-resource Veps and Karelian languages.

Algorithms for assigning part of speech tags to words and grammatical properties to words are presented in the article. These algorithms use our morphological dictionaries, where the lemma, part of speech and a set of grammatical features (gramset) are known for each word form.

The algorithms are based on the analogy hypothesis that words with the same suffixes are likely to have the same inflectional models, the same part of speech and gramset.

The accuracy of these algorithms were evaluated and compared. 313 thousand Vepsian and 66 thousand Karelian words were used to verify the accuracy of these algorithms. The special functions were designed to assess the quality of results of the developed algorithms.

71,091 "word – part of speech" pairs for the Karelian Proper supradialect and 399,260 "word – part of speech" pairs for the Veps language have been used in the experiments to evaluate algorithms. 86.77% of Karelian words and 92.38% of Vepsian words were assigned a correct part of speech.

73,395 "word – gramset" pairs for the Karelian Proper supradialect and 452,790 "word – gramset" pairs for the Veps language have been used in the experiments to evaluate algorithms. 90.7% of Karelian words and 95.3% of Vepsian words were assigned a correct gramset by our algorithm.

If you need only one correct answer, then all three of developed algorithms are not very useful. But in our case, the task is to get an ordered list of the parts of speech and gramsets for a word and to offer this list to an expert. Then the expert selects the correct part of speech and gramset from the list and assigns to the word. This is a semi-automatic tagging of the texts. Thus, these algorithms are useful for our corpus.

Acknowledgement. The study was carried out under the state order of the Institute of Applied Mathematical Research and the Institute of Language, Literature and History of Karelian Research Centre of the Russian Academy of Sciences.

References

1. Belonogov, G.G., Kalinin, Y.P., Khoroshilov, A.A.: Computer Linguistics and Advanced Information Technologies: Theory and Practice of Building Systems for Automatic Processing of Text Information. Russian World, Moscow (2004). (in Russian)
2. Hämäläinen, M.: UralicNLP: an NLP library for Uralic languages. J. Open Source Softw. **4**(37), 1345 (2019). https://doi.org/10.21105/joss.01345
3. Moshagen, S., Rueter, J., Pirinen, T., Trosterud, T., Tyers, F.M.: Open-source infrastructures for collaborative work on under-resourced languages. In: Collaboration and Computing for Under-Resourced Languages in the Linked Open Data Era, Reykjavík, Iceland, pp. 71–77 (2014)
4. Pirinen, T.A., Trosterud, T., Tyers, F.M., Vincze, V., Simon, E., Rueter, J.: Foreword to the special issue on Uralic languages. Northern Eur. J. Lang. Technol. **4**(1), 1–9 (2016). https://doi.org/10.3384/nejlt.2000-1533.1641

Extrinsic Evaluation of Cross-Lingual Embeddings on the Patent Classification Task

Anastasiia Ryzhova[1(✉)] and Ilya Sochenkov[1,2]

[1] Federal Research Center "Computer Science and Control" of Russian Academy
of Sciences, 44-2 Vavilov Str., Moscow 119333, Russia
ryzhova@tesyan.ru
[2] HSE University, Moscow, Russia
isochenkov@hse.ru

Abstract. In this article we compare the quality of various cross-lingual embeddings on the cross-lingual text classification problem and explore the possibility of transferring knowledge between languages. We consider Multilingual Unsupervised and Supervised Embeddings (MUSE), multilingual BERT embeddings, XLM-RoBERTa (XLM-R) model embeddings, and Language-Agnostic Sentence Representations (LASER). Various classification algorithms use them as inputs for solving the task of the patent categorization. It is a zero-shot cross-lingual classification task since the training and the validation sets include the English texts, and the test set consists of documents in Russian.

Keywords: Patent classification · Cross-lingual embeddings ·
Zero-shot cross-lingual classification · MUSE embeddings · LASER
embeddings · Multilingual Bert · XLM-RoBERTa

1 Introduction

The patent classification is a challenging and time-consuming task, and it is mainly carried out by specialists manually. At this moment, there already exist large databases of patent documents, which allows creating more and more accurate methods of automatic classification of patent documents.

The patent is a legal document that allows its holder to have an exclusive right to control the use of the new invention. It stops other people from selling this invention without authorization. The subject of a patent can be any new, non-obvious invention. It can cover new mechanical devices, methods, and processes, chemical compositions or compounds, computer programs, etc.

1.1 IPC Taxonomy

The International Patent Classification (IPC) is a hierarchical classification of patent documents. This universal system was created to simplify the search for

© Springer Nature Switzerland AG 2021
A. Sychev et al. (Eds.): DAMDID/RCDL 2020, CCIS 1427, pp. 178–190, 2021.
https://doi.org/10.1007/978-3-030-81200-3_13

patent documents in different languages. Each document is assigned one or more indices.

There exist eight main sections, which are denoted by letters of the Latin alphabet from A to H:

A: Human Necessities
B: Performing Operations, Transporting
C: Chemistry, Metallurgy
D: Textiles, Paper
E: Fixed Constructions
F: Mechanical Engineering, Lighting, Heating, Weapons
G: Physics
H: Electricity

It is the highest level of the hierarchy, the next levels, such as class, subclass, group, and subgroup, provide more in-depth specificity (Fig. 1). Sometimes it is not easy to define one IPC code for the invention, and the expert assigns two or even more IPC classifications.

Subdivision	Example of an IPC code	
	Symbol	Title
Section	G	Physics
Class	G06	Computing; calculating; counting
Sub-class	G06Q	Data processing systems or methods, specially adapted for administrative, commercial, financial, managerial, supervisory or forecasting purposes
Main group	G06Q 20/00	Payment architectures, schemes or protocols
Sub group	G06Q 20/10	...specially adapted for electronic funds transfer [EFT] systems

Fig. 1. Example of an IPC code, from [13]

1.2   Structure of Patent Documents

Patent documents have a specific structure. It includes claims, detailed description, drawings, background, abstract, and summary. The claims are the heart of a patent document and contain the description of novelty. There describes what the patent agent wants to protect. The description section includes a detailed explanation of the invention. The drawings contain visualizations with short descriptions of the presented figures. In the background, the patent agent should describe the prior art of the invention. The summary is usually the final part of the patent document, which summarizes all described earlier. In our work, we consider the abstract, the short description of the invention, where the essence of the invention must be described very clearly and use a minimum number of words.

2   Related Work

2.1   Patent Classification

The automatic classification of patent documents is a non-trivial task. A lot of scientific research is devoted to this problem. Fall et al. [7] applied various machine learning algorithms, including the nearest neighbor method (KNN), the support vector machine (SVM), and naive Bayes to automatically classify patent documents into index classes and subclasses in English. They analyzed which parts of the patent document most affect the quality of classification, and proposed various approaches to classifying documents into several classes (multilabel classification).

In their next paper [8], the same researchers performed a similar analysis for the automatic classification of documents in German, comparing the results with those obtained earlier in English.

In addition to various machine learning methods, scientists took into account the structure of the patent document as well. Searching for similar documents, Jae-HoKim et al. [14] compared only certain semantic elements (for example, application field), rather than full texts. They clustered these semantic elements, and then for the document class identification, they used the K nearest neighbor classification algorithm. Lim et al. [16] showed that the quality of classification is strongly influenced by structural fields such as technical field and background field.

Most of the articles are focused on patent classification at the level of classes and subclasses of indices. Chen et al. [2] were among the first who classified patent documents to the level of subgroups. They obtained an accuracy which equaled 36.07%. A detailed comparison of various patent classification researches can be found in [9].

In [6] the researchers performed the automated patent classification for four languages: English, French, Russian and German. They even developed the system interface prototype for the automated categorization by abstract or the full text of patent document.

In this paper, the improvement of existing patent classification methods was not the primary goal. The aim was to compare the quality of cross-language embeddings (extrinsic evaluation) on zero-shot cross-lingual classification task, using English and Russian patent documents.

2.2   Cross-Lingual Embeddings

Currently, word vector representations are an essential tool for solving natural language processing tasks. Cross-lingual word embeddings are in a particular interest in two reasons. Firstly, their primary purpose is to compare the meanings of words in different languages. Secondly, the universal representation of words in different languages will help to analyze and solve the NLP tasks in low-resource languages by transferring the knowledge from rich-resource ones.

Scientists noticed that most of the geometric relationships between words are preserved in different languages, making it possible to build mapping matrices from the vector space of words in one language to the vector space of words in another language. Mikolov et al. [17] selected 5,000 of the most frequently used words in the source language, translated them into the target language, and then minimized the distance between the word-translation pairs with stochastic gradient descent. In other words, they searched for a mapping matrix W

$$\min_W \sum_{i=1}^{n} |Wx_i - z_i|,$$

where x_i is a monolingual representation of source word w_i, and z_i is its translation. Since 2013, the researches presented many ways of cross-lingual embeddings constructions, which can be found in detail in [19]. Further, we consider the MUSE, LASER, XLM-RoBERTa, and MBERT model embeddings that we compared in this study.

Multilingual unsupervised and supervised embeddings, MUSE, [4,15] were presented by Facebook AI researchers. The starting point was fasttext word embeddings, which were later embedded in the common space in two ways, by supervised and unsupervised approaches. In the first case, the researchers used a bilingual dictionary and constructed the mapping from source to target space using Procrustal alignment [20]. In the second case, no parallel data was used, and the adversarial training was applied to get the mapping matrix. Let's assume that $X = \{x_1, x_2, \ldots x_n\}$ is source embeddings, and $Y = \{y_1, y_2, \ldots y_m\}$ is target embeddings, W is the desired mapping matrix. The idea of the method is that there is a competition between two competitors [10]: one model, the discriminator, tries to distinguish elements randomly selected from WX from Y, while W is trained to make WX and Y more similar.

LASER embeddings [1] were also released in Facebook AI. These are Language-Agnostic SEntence Representations, universal embeddings that do not depend on the language. Each sentence is represented as a point in a multidimensional space. The main goal is to get points so that the same sentences in different languages are as close to each other as possible. Such representations can be considered as a universal language in the semantic vector space. To get such vector representations, researchers suggest using the encoder-decoder neural network architecture. Encoder with a shared BPE vocabulary for all languages consists of a bidirectional long-term memory network (BiLSTM) with five layers. The sentence embedding is obtained via max-pooling over the last states of the LSTM layers and has a dimension of 1024. Then this vector, language ID vector, and input vector are passed to the decoder as input. The encoder does not have any information about language, that allows it to generate output embeddings that do not depend on the language. The network was trained on a parallel corpus of 223 million sentences in English and Spanish. Then training was conducted for 21 languages, which improved the quality of the model. Byte pair encoding (BPE) vocabulary, constructed on the concatenation of the languages, even allowed obtaining generic characteristics of language families.

The Multilingual Bert model does not differ a lot from the usual Bert model [5]. It is a Transformer model [21] with twelve layers. One exception is that it was pre-trained on Wikipedia texts from the top 104 languages with a shared word piece vocabulary and no parallel cross-lingual data was used.

The XLM-RoBERTa model [3] is a new state-of-the art multilingual model released in November 2019. It has a transformer architecture as well. The most significant update that this model was trained on a massive amount of data, namely, on 2.5T of filtered CommonCrawl data. The XLM-R outperforms the MBert on classification by up to 23% accuracy on low-resource languages, and even shows the competitive results with state-of-the-art monolingual models.

3 Data

3.1 Data Description

English is a widely used language, and it is a common approach to use it for training models and then to transfer the knowledge to lower resourced languages.

We performed the automatic classification at the subclass level (the first four digits of the index) and selected the documents that belonged to only one subclass. There are 80,948 patent documents in the English language, which we divided into training and validation sets in the ratio of 85%/15%. These sets include patent documents of 225 different subclasses. As the test set, we consider 10,150 texts in Russian of the same subclasses. Table 1 shows detailed statistics. The last three columns correspond to the median, minimum, and maximum number of documents in one subclass.

Table 1. Data statistics

	Total	Median	Minimum	Maximum
Train set	68805	294	85	510
Validation set	12143	52	15	90
Test set	10150	43	13	75

The "abstract" section of a patent document contains about one hundred words on average. Figure 2 shows the length distributions of the training, validation, and test sets.

An additional complication in patent classification is that some subclasses contain a large number of uncommon words (for example, in the section "Chemistry", which includes descriptions of chemical elements). After calculating the average number of unique concepts for each subclass in our training sample, we made sure that our data does not have a broad diversity in the size of dictionaries for subclasses. Each abstract text contains about fifty unique words.

To make sure that the texts of the same subclasses in English and Russian are semantically close, we constructed a top keyword list for each subclass using

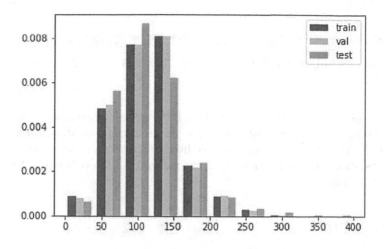

Fig. 2. Lengths distribution

lemmatized documents. We computed the importance of the word by tf-idf measure, separately for train English texts and test Russian texts. For example, we show the most important concepts for some subclasses (the terms are listed in descending order of significance) in Fig. 3.

A01M		C12N		G01F	
Top English	**Top Russian**	**Top English**	**Top Russian**	**Top English**	**Top Russian**
decoy	опрыскиватель	cell	клетка	flowmeter	измерение
trap	насекомое	gene	днк	meter	жидкость
insect	ловушка	protein	рекомбинантный	measurement	измерительный
bait	распылитель	culture	биотехнология	transducer	расходомер
termite	труп	dna	штамм	ultrasonic	расход
pest	уничтожение	nucleic	ген	conduit	уровень
rodent	вредитель	recombinant	вирус	mass	дозатор
animal	опрыскивание	polypeptide	полипептид	container	датчик
attractant	гелевый	acid	растение	coriolis	поток
waterfowl	приманка	expression	белок	dispense	ёмкость

Fig. 3. Top keywords of three subclasses

3.2 Data Preprocessing

First of all, for all experiments we converted all the texts to the lower case. Applying MUSE embeddings, we deleted the numbers and all punctuation as well. In other cases, we kept the points and commas, since other punctuation is

Table 2. How MUSE cover documents vocabulary

	Train texts (eng)	Val texts (eng)	Test texts (rus)
mean ± std, %	98.77 ± 2.25	98.78 ± 2.23	88.62 ± 5.63

not widely used in the patent documents. Before some experiments, we lemmatized the texts. For the English language, we used spacy module, for Russian - pymystem3. Also, we wanted to compare how the results would change after test texts translation in English. For this reason, we translated Russian texts with Yandex Translator.

4   Experiments

4.1   Multilingual Unsupervised and Supervised Embeddings

In this section we describe the experiments with MUSE embeddings. The maximum size of the embeddings vocabulary equals 200 thousand for both Russian and English languages. For each text, we computed the ratio of words that occurred in the corresponding vocabulary. The results are shown in Table 2.

As one can see from the table, the number of present words in MUSE vocabulary in Russian texts differs from the number of present words in English on ten percent. We computed the occurrences of lemmatized words in the MUSE vocabulary as well. For train and validation sets the percentage did not differ a lot, but for the test set the mean value increased by one percent. For this reason we decided to use the lemmatized texts.

We started to hold our experiments with machine learning models, such as logistic regression and the SVM classifier. As inputs, we used the average sum of all word embeddings in the text. To the same input matrices, we also applied the Feed Forward Neural Network with two layers and ReLU activation. In the future, it will also be interesting to apply more time-consuming methods, e.g., xgboost classifier. Then we trained the bidirectional LSTM models with simple dot product attention and the convolutional neural networks with three convolutional layers and adaptive max pooling. The parameters of both models (hidden size, dropout, and the number of layers in LSTM; output size, kernel sizes of convolutions, dropout) were chosen by Random search. The final BiLSTM model has the following parameters: hidden size = 128, 1 layer, dropout = 0. The CNN model consists of three convolutional layers with kernel sizes = $[2, 3, 4]$, dropout = 0.5, output size = 328. We do not have a massive imbalance in our data, and in this task it is not so significant to have high accuracy on small classes, so we decided to use the f1-micro and f1-weighted scores as evaluation metrics. The results are presented in Table 3. The first value corresponds to f1-micro, the second to the f1-weighted score. "Test translated" is the Russian test documents which were translated with the Yandex translator.

Table 3. Results of experiments with MUSE embeddings, f1-micro/f1-weighted scores

Model	Train	Validation	Test	Test translated
SVM	48.63/47.53	44.92/43.52	12.81/11.97	38.70/37.80
Logistic regression	44.40/42.80	42.10/40.24	19.87/17.27	37.54/34.98
FFNN	43.92/42.35	41.92/40.17	17.48/15.26	38.56/36.28
BiLSTM with attention	55.33/54.42	51.54/50.63	**25.36/23.41**	47.75/46.14
CNN	60.50/59.89	51.67/50.98	22.62/20.97	44.95/43.88

The best results were achieved with the BiLSTM model, but the scores on the test set are much lower, than on validation set and on the translated test texts.

When we use the average of all word embeddings, we can lose the information, since some terms can make a more exceptional contribution to subclass identification. For this reason, we conducted experiments on classification by weighted top keywords vectors. We fitted one TfIdfvectorizer with the train texts and the separate one with the test texts. Then the top-15 keywords with their tf-idf values were extracted from each document. Then the top-15 keywords with their tf-idf values were extracted from each document. We choose the number 15 since there were too few words in the MUSE vocabulary for some documents with the lower number. Then we computed the weighted average of keywords embeddings and applied the logistic regression for classification. For test documents, we used two options. In one case, we took the word embeddings from MUSE Russian vocabulary. In another case, we search for the most similar English words (TOP-1, TOP-3, and TOP-5) to the Russian one, and then used the average vector for these English terms as a vector for one Russian keyword. The results (Table 4) show that the second option can improve classification accuracy.

Table 4. The classification results by top keywords, logistic regression model

Train texts (eng)	Val texts (eng)	Test texts (rus)	Test texts Top-1 similar	Test texts Top-3 similar	Test texts Top-5 similar
44.79/43.33	42.47/40.98	23.87/21.9	**25.98/23.82**	25.73/23.63	25.03/23.03

4.2 Language-Agnostic Sentence Representations

LASER embeddings are sentence representations. We vectorized our documents in two ways. One approach was to embed each sentence of text separately and then to take the average of these sentence embeddings. Another method was to apply the LASER encoder model to the whole document at once since we deal with short texts. We obtained the sentence embeddings of the first thousand Russian documents and their translations using these two approaches. Then in both cases, we computed the cosine similarity between the vectors of texts and the vectors of their translations. It turned out that in a second way the numbers were higher in 85% of cases, so we chose this approach.

As earlier, the proposed algorithms consist of training three models: logistic regression, Feed-forward Neural Network (FFNN) with one hidden layer of size 256, and SVM classifier. The searching for the best parameters in logistic regression and FFNN models was performed with the Optuna Python module. For example, in the FFNN model, the objective function, f1-micro score on the validation set, was maximized by searching for the best parameters among the number of layers, dropout, hidden sizes, and learning rate. The parallel coordinate plot of this search is presented in Fig. 4.

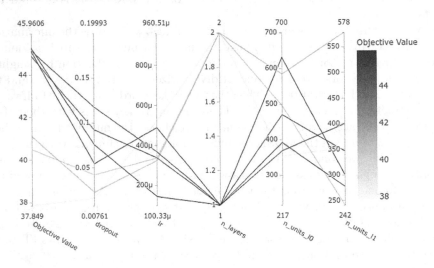

Fig. 4. Search for hyperparameters of FFNN model

Table 5 shows the results of training models with best parameters. Compared to previous experiments, we managed to increase the f1-micro score on test documents by six percent.

4.3   Multilingual BERT and XLM-RoBERTa

At the last stage, we fine-tuned the Multilingual Bert and XLM-RoBERTa models on our patent classification task. For classification, we use the linear layer on the top of the last pooling layer. In MBert case, for better performance, we froze the embedding layer to make the model less tuned to the English language. The models were trained with the following parameters: max_seq_length = 256, batch size = 8. From Table 6, one can see that these models achieve the best results on the validation and test translated datasets, and XLM-R embeddings outperform the other models on test documents.

Table 5. Results of experiments with LASER embeddings, f1-micro/f1-weighted scores

Model	Train	Validation	Test	Test translated
SVM	56.33/55.70	43.57/42.05	28.78/28.34	38.74/37.05
Logistic regression	54.76/53.85	43.87/42.16	31.72/29.56	39.69/37.84
FFNN	56.46/55.49	45.22/44.01	**32.82/31.14**	40.38/38.69

Table 6. F1-micro/f1-weighted score after MBert and XLM-Roberta models fine-tuning

Model	Train	Validation	Test	Test translated
MBert	71.96/70.50	**61.62/60.15**	25.05/21.81	**59.52/57.68**
XLM-Roberta	70.43/68.84	61.44/59.84	**41.06/38.69**	**59.76/57.82**

5 Results Discussion

Table 7 presents the summary, the highest f1-micro and f1-weighted scores which were achieved with different cross-lingual embeddings.

Table 7. The best experiments with different cross-lingual embeddings, summary table

Model	Train	Validation	Test	Test translated
MUSE + BiLSTM with attention	55.33/54.42	51.54/50.63	25.36/23.41	47.75/46.14
LASER + FFNN	56.46/55.49	45.22/44.01	32.82/31.14	40.38/38.69
MBert	71.96/70.50	**61.62/60.15**	25.05/21.81	**59.52/57.68**
XLM-Roberta	70.43/68.84	61.44/59.84	**41.06/38.69**	**59.76/57.82**

In this section, we want to discuss the obtained results.

(1) What Are the Disadvantages of MUSE Embeddings, Why Do We Have Such a Big Gap Classifying the Documents on Different Languages? We think that one of the reasons is the disbalance between the presence of the words in MUSE vocabulary. It can be crucial when the topic keyword appears in the MUSE vocabulary of English terms, but we do not find it among the Russian words. Also, we noticed that the more the models overfit (the greater the difference between train and validation scores), the lower the test performance. Therefore, it is important to set strict early stoppings criterion in the training process. Moreover, the disadvantage of MUSE embeddings is that the vector representation of the word does not depend on the context.

(2) How Can the Performance of LASER Embeddings Be Improved? In [11] the researchers proposed an additional improvement to the LASER embeddings. They developed Emu system, which can semantically enhance these embeddings. It includes multilingual adversarial training and uses only mono-lingual labeled data. There are three main parts in the model (Fig. 5): the mul-tilingual encoder, the semantic classifier, which groups the sentences with equal intent, and the language discriminator, which tries to distinguish the languages and mislead the other two parts of the system. The experimental results showed that EMU outperformed the LASER embeddings.

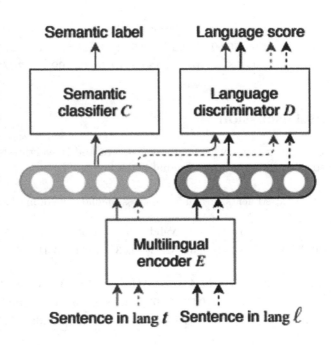

Fig. 5. Emu architecture

(3) **We Have a Big Gap Between Validation and Test Scores After Experiments with MBert Model. Why?** In paper [18], the researchers investigated how the MBert model performs the transfer between languages. The model was fine-tuned on specific tasks on one language and then evaluated in another language. They showed that it is possible to achieve good results even with zero lexical overlaps, it is also confirmed by experiments in [12]. On the contrary, the different word ordering in languages is the reason for the significant drop in the performance, and it clarifies our obtained scores.

(4) **What Do the Results on Translated Texts Tell Us?** There is a big difference between the scores on test data in Russian and translated test texts in English. It is fast to apply the cross-lingual embeddings and train the models,

but their performance is not high compared to the performance on the validation. We have three hypotheses why it is so:

- The Yandex Translator has a high quality of the translation from Russian to English language, even if it is a statistical machine translation. (We used the free version of the Translator which does not leverage the neural networks). It can be checked by computation the BLEU metric.
- The models, which take into account the sequences of words, such as BiLSTM models, can not transfer this knowledge from English to the Russian language, since they have a different structure.
- The patent classification task has domain-specific vocabulary, and it turns out that the cross-lingual embeddings do not catch well the semantic similarity between some special terms. The XLM-RoBERTa model was trained on more than two terabytes of filtered CommonCrawl data, and it is one of the reasons this model has a minimal gap between f1-micro scores on test and test translated data.

6   Conclusions

In our research, we performed an extrinsic evaluation of four types of cross-lingual vector representations: MUSE, LASER, MBert, and XLM-RoBERTa embeddings. We can conclude that the XLM-R model embeddings outperform other approaches and show a better ability to transfer knowledge between languages. However, in practical applications, combining this model with the Yandex translator can be more useful, even though this method is more time-consuming.

In future work, the results can be improved in different ways. Firstly, we can extract more information from the patent documents, adding other fields of the patent documents. Secondly, we can use MUSE embeddings as inputs to models with LSTM-CNN architectures or fine-tune MBert and XLM-RoBERTa models considering the RNN layer on top of the pooled output of Bert/XLM-RoBERTa. Finally, it seems fruitful to combine models with different cross-lingual embeddings in the neural network ensemble and check the results.

Also, it would be interesting to compare the results using test documents in other languages.

Python source code is available online[1].

Acknowledgments. The reported study was funded by RFBR according to the research projects No 18-37-20017 & No 18-29-03187. This research is also partially supported by the Ministry of Science and Higher Education of the Russian Federation according to the agreement between the Lomonosov Moscow State University and the Foundation of project support of the National Technology Initiative No 13/1251/2018 dated 11.12.2018 within the Research Program "Center of Big Data Storage and Analysis" of the National Technology Initiative Competence Center (project "Text mining tools for big data").

---

[1] https://github.com/ryzhik22/Cross-lingual-embeddings-evaluation.

References

1. Artetxe, M., Schwenk, H.: Massively multilingual sentence embeddings for zero-shot cross-lingual transfer and beyond (2018)
2. Chen, Y.L., Chang, Y.C.: A three-phase method for patent classification. Inf. Process. Manag. **48**, 1017–1030 (2012). https://doi.org/10.1016/j.ipm.2011.11.001
3. Conneau, A., et al.: Unsupervised cross-lingual representation learning at scale (2019)
4. Lample, G., et al.: Word translation without parallel data. In: International Conference on Learning Representations (2018)
5. Devlin, J., Chang, M.W., Lee, K., Toutanova, K.: Bert: pre-training of deep bidirectional transformers for language understanding (2018)
6. Fall, C., Benzineb, K., Guyot, J., Törcsvári, A., Fiévet, P.: Computer-assisted categorization of patent documents in the international patent classification (2003)
7. Fall, C., Törcsvári, A., Benzineb, K., Karetka, G.: Automated categorization in the international patent classification. SIGIR Forum **37**, 10–25 (2003). https://doi.org/10.1145/945546.945547
8. Fall, C., Törcsvári, A., Fiévet, P., Karetka, G.: Automated categorization of German-language patent documents. Expert Syst. Appl. **26**, 269–277 (2004). https://doi.org/10.1016/S0957-4174(03)00141-6
9. Gomez, J.C., Moens, M.-F.: A survey of automated hierarchical classification of patents. In: Paltoglou, G., Loizides, F., Hansen, P. (eds.) Professional Search in the Modern World. LNCS, vol. 8830, pp. 215–249. Springer, Cham (2014). https://doi.org/10.1007/978-3-319-12511-4_11
10. Goodfellow, I.: Nips 2016 tutorial: generative adversarial networks (2016)
11. Hirota, W., Suhara, Y., Golshan, B., Tan, W.C.: Emu: enhancing multilingual sentence embeddings with semantic specialization (2019)
12. Wang, Z., Mayhew, S., Roth, D.: Cross-lingual ability of multilingual BERT: an empirical study (2019)
13. Kapoor, R.: Intellectual property and appropriability regime of innovation in financial services, p. 33 (2014)
14. Kim, J.H., Choi, K.S.: Patent document categorization based on semantic structural information. Inf. Process. Manag. **43**, 1200–1215 (2007). https://doi.org/10.1016/j.ipm.2007.02.002
15. Lample, G., et al.: Unsupervised machine translation using monolingual corpora only. In: International Conference on Learning Representations (2018)
16. Lim, S., Kwon, Y.J.: IPC multi-label classification applying the characteristics of patent documents. In: Park, J.J.J.H., Pan, Y., Yi, G., Loia, V. (eds.) CSA/CUTE/UCAWSN -2016. LNEE, vol. 421, pp. 166–172. Springer, Singapore (2017). https://doi.org/10.1007/978-981-10-3023-9_27
17. Mikolov, T., Le, Q., Sutskever, I.: Exploiting similarities among languages for machine translation (2013)
18. Pires, T., Schlinger, E., Garrette, D.: How multilingual is multilingual BERT? pp. 4996–5001 (2019). https://doi.org/10.18653/v1/P19-1493
19. Ruder, S.: A survey of cross-lingual embedding models (2017)
20. Schönemann, P.: A generalized solution of the orthogonal procrustes problem. Psychometrika **31**, 1–10 (1966). https://doi.org/10.1007/BF02289451
21. Vaswani, A., et al.: Attention is all you need (2017)

An Approach to Extracting Ontology Concepts from Requirements

Marina Murtazina$^{(\boxtimes)}$ (iD) and Tatiana Avdeenko (iD)

Novosibirsk State Technical University, Novosibirsk 630073, Russia
murtazina@corp.nstu.ru

Abstract. The paper proposes an approach to extracting ontology concepts from the results of automatic processing the software requirements written in natural Russian-language. Relevance of developing an approach to automating the process of extraction of the ontology concepts from the requirements written in natural language is due to the necessity to maintain the requirements specification in a consistent state under conditions of business environment variability. First we discuss the advantages of the ontology-based approach to the requirements engineering as a stage of the software engineering process. The main attention is focused on the possibilities of presenting knowledge about the software application domain and requirements specification in the form of ontologies. We analyzed possibilities of automatic Russian text processing tools for extracting ontology concepts and consider such tools as ETAP-4, MaltParser and UDPipe. The choice of UDPipe as a tool for automatic processing the requirements texts is explained by the best results that it showed when analyzing texts such as software requirements. Then we describe the main classes, object properties and data properties of the proposed requirements ontology and the application domain ontology. An additional advantage of the proposed approach, increasing its practical utility, is building the system of production rules by which the analysis of the results of the automatic processing of the requirements texts obtained using the UDPipe tool.

Keywords: Requirements engineering · Ontology · Automatic text processing

1 Introduction

The most critical subprocess of the software engineering process, the results of which affect the success of the software development project, is the requirements engineering process. Initially the requirements are usually recorded in natural language. A requirements array accumulates and changes during the implementation of the software development project. It is important to ensure the completeness of the description, the uniqueness and traceability of the requirements, and to correctly present the requirements in the form of consistent set of the requirements. The criterion of consistency is fulilled, if the set of the requirements

© Springer Nature Switzerland AG 2021
A. Sychev et al. (Eds.): DAMDID/RCDL 2020, CCIS 1427, pp. 191–203, 2021.
https://doi.org/10.1007/978-3-030-81200-3_14

does not contain conflicting requirements, the requirements are not duplicated, and the same terms are used for the same elements in all requirements. The trend of research in the field of improving the developing requirements process is presentation of the requirements as facts from the domain model. Numerous studies have proven the effectiveness of using ontologies for these purposes [1–3]. The variability of requirements on the part of stakeholders, the need for quick comparison of the requirements texts, the need to extract the concepts of the application domain area for requirements analysis determine the topicality of developing approaches to automating the extraction of ontology concepts from texts containing requirements.

The paper is organized as follows. Section 1 substantiates the relevance of the research topic. Section 2 analyzes related works. Section 3 studies the possibilities of automatic Russian text processing to extracting requirements ontology concepts. Section 4 proposes an approach to converting textual requirements to ontology. Section 5 summarizes the work.

2   Related Works

One of the first works in the field of textual requirements analysis based on ontologies was the paper [4], in which a method was proposed for determining the correspondence between the software requirements specification and the domain ontology. The system of ontologies proposed in [4] included a thesaurus and inference rules. At the same time, the authors of this approach did not use NLP techniques, and the technique they proposed was defined as Lightweight Semantic Processing. Subsequently, the automatic construction of UML diagrams from textual requirements became an active research sphere in the field of text requirements analysis [5,6]. At the same time, NLP methods and domain ontologies were widely used. The manual construction of domain ontologies required a lot of effort. In this connection, scientific interest arose in the development of methods for enriching the software product domain ontologies with new concepts extracted from textual requirements.

The issues of extracting ontology concepts from the requirements and checking the consistency of the requirements specification based on natural language processing methods are considered in [7,8]. In [9], based on the Stanford CoreNLP tools, an approach to processing requirements written in a natural language is proposed, designed to transform sets of heterogeneous requirements in an ontology. In [10], an approach is proposed to automatically construction of the ontology from a set of user stories. The spaCy library is used to process text in natural English. In [11], a set of rules is proposed for constructing the domain ontology based on content analysis of text resources. For automatic text processing, the AOT RML toolkit is used.

In [12], an approach is proposed for comparing the point of stakeholders view on the subsystems of a specific information system. To compare textual requirements, the ontology concepts are extracted using the system of logical rules in a pre-prepared structure of the ontology. Two textual descriptions of the system are extracted into two ontologies. The concepts of the obtained ontologies are compared with each other using the N-gram method, as well as linguistic ontologies. As a result of comparison between the concepts of two textual descriptions of the requirements, semantic relations were established. Stakeholder requirements were presented in natural English. The TreeTagger text morphological markup tool and the Stanford Parser tool for constructing syntax dependency trees were used to process the textual requirements.

Integration of data containing user requirements produced in different branches of the same company or companies from the same industry requires the development of methods for matching the terms used by various employees. The detection of semantic conflicts in user requirements is discussed in [13]. An ontology-based approach that combines vocabularies and formalisms to express user requirements, such as the UML use cases, the goal-oriented requirements model and the Merise method, is given in [14].

The analysis of publications on the research topic allows us to conclude that one of the promising areas of research is to extract ontology concepts from textual requirements. Automatic processing of textual requirements is especially in demand as a part of conflict analysis methods and the search for similar requirements. In these methods, requirements are represented in the form of ontology concepts, and then they are compared according to logical rules that define the relationships between actors, actions, and objects. Nevertheless, the automatic filling of the requirements ontologies from the texts of the requirements is currently still a poorly covered topic. The vast majority of scientific works proposed methods for working with English-language requirements. This paper explores the possibilities of extracting the concepts of the requirement ontology from the results of the automatic processing of Russian-language requirements.

3   Analysis of Automatic Russian Text Processing Tools Possibilities for Extracting Ontology Concepts

The use of natural language processing methods to extract requirements ontology concepts involves syntactically sequential constructing sentences with the requirements. Sentences with the requirements should begin with the actor, followed by a verb expressing the action. The action should be followed by the object to which the subjects action is directed [15]. During the research, the ETAP-4, MaltParser and UDPipe tools were considered for marking up the requirements presented in natural Russian-language. ETAP-4 is a linguistic analyzer based on the rules of linguistic analysis and synthesis according to the Meaning-Text theory with some elements of statistics and machine learning. Currently, this linguistic processor is licensed under CC BY-NC-ND 4.0. An example of the dependency tree for textual requirement builted by WinEtap 4.1.9.3 is shown in Fig. 1.

MaltParser is a language-independent tool for working with dependency trees. This text analyzer enables to train models according to annotated corpus and build dependency trees for new data, based on a ready-made language

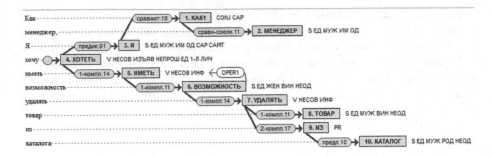

Fig. 1. A dependency tree built by WinEtap-4.

model. Since MaltParser does not perform tokenization, morphological tagging and lemmatization, it is necessary to convert text data into one of the input formats before using it. At the same time, for morphological tagging, a tool should be used that supports the same model of morphological features of speech parts as the language model used by MaltParser. Files in the following formats CoNLL-X, CoNLL-U or MaltTAB are used as the input to the parser. Dependency analysis results are also saved in one of these three formats. There is a open access Russian language model developed by S. Sharov et al. [16,17] for MaltParser. It is worth noting that in the language model S. Sharov et al. the same set of dependency relationships is used as in the ETAP-4 system. An example of the result of the analysis of dependencies in a textual requirement built by MaltParser (version 1.9.2) is shown in Fig. 2.

```
pars_out.conll
1     1——Как——как——C——C——C——5——вводн——,——_——LF
2     2——менеджер——менеджер——N——N——Ncmsny——1——сравн-союзн——_——LF
3     3——,——,——,——,——,——2——PUNC——,——_——LF
4     4——Я——я——P——P——P-1-snn——5——предик——_——_——LF
5     5——хочу——хотеть——V——V——Vmip1s-a-e——0——ROOT——_——_——LF
6     6——удалять——удалять——V——V——Vmn----a-e——5——1-компл——_——_——LF
7     7——товар——товар——N——N——Ncmsan——6——1-компл——_——_——LF
8     8——из——из——S——S——Sp-g——6——2-компл——_——_——LF
9     9——каталога——каталог——N——N——Ncmsgn——8——предл——_——_——LF
10    10——.——.——S——S——SENT——9——PUNC——_——_——LF
11    LF
```

Fig. 2. A dependency tree built by MaltParser (in CoNLL-U format).

The results of parsers are almost the same. Therefore, the convenience of integration with other software systems was a decisive factor for choosing between these two tools for research purposes. Since MaltParser is a java application, its use becomes preferable because the execution of the application does not depend on the operating system, whereas ETAP-4 is Windows-based software.

As a part of this research, a language-independent TreeTagger tool (version 3.2.1) was used for morphological tagging and lemmatization of the text. For automatic processing of simple text files (files containing only text), a set of scripts in Python (version 3.7.0) and a batch file for execution of scripts under the Windows operating system were developed (a similar shell script can be made for Linux family of OSs). The developed program performs the following actions on the source data of a simple text file:

1. text segmentation (division into sentences),
2. tokenization of sentences (division of each sentence into words, including punctuation in the sentence),
3. morphological analysis using TreeTagger,
4. analysis of syntactic dependencies using MaltParser.

The next tool, that MaltParser is compared to be, is trainable pipeline UDPipe. This tool performs tokenization, tagging, lemmatization and dependency parsing of CoNLL-U. UDPipe is licensed under the Mozilla Public License 2.0. Figure 3 shows a dependency tree in CoNLL-U format of the sentence "As a manager, I want to remove goods from the catalog".

Id	Form	Lemma	UPosTag	XPosTag	Feats	Head	DepRel	Deps	Misc
# newdoc									
# newpar									
# sent_id = 1									
# text = Как менеджер, Я хочу удалять товар из каталога.									
1	Как	как	SCONJ	_	_	2	mark	_	_
2	менеджер	менеджер	NOUN	_	Animacy=Anim|Case=Nom|Gender=Masc|Number=Sing	5	parataxis	_	SpaceAfter=No
3	,	,	PUNCT	_	_	2	punct	_	_
4	Я	я	PRON	_	Case=Nom|Number=Sing|Person=1	5	nsubj	_	_
5	хочу	хотеть	VERB	_	Aspect=Imp|Mood=Ind|Number=Sing|Person=1|Tense=Pres|VerbForm=Fin|Voice=Act	0	root	_	_
6	удалять	удалять	VERB	_	Aspect=Imp|VerbForm=Inf|Voice=Act	5	xcomp	_	_
7	товар	товар	NOUN	_	Animacy=Inan|Case=Acc|Gender=Masc|Number=Sing	6	obj	_	_
8	из	из	ADP	_	_	9	case	_	_
9	каталога	каталог	NOUN	_	Animacy=Inan|Case=Gen|Gender=Masc|Number=Sing	6	obl	_	SpaceAfter=No
10	.	.	PUNCT	_	_	5	punct	_	SpaceAfter=No

Fig. 3. A dependency tree built by UDPipe.

As a part of the study of the possibilities of using the MaltParser and UDPipe tools for the purposes of this work, Russian language models were trained on the UD_Russian-SynTagRus case (version 2.0-170801). After training both tools on the same data, a qualitative comparison of the results was carried out. It is worth noting that in MaltParser results for complex sentences sometimes incorrect results. MaltParser does not always correctly assign the tag "root": it can set this tag for a verb, it can for a noun (if there is a verb that must be defined as a root of the syntax dependency tree), it can set a tag "root" for several words. An example of the dependency tree with two root tags built by MaltParser is shown in Fig. 4.

Fig. 4. An example of a dependency tree with two roots (built by MaltParser).

Possibility of obtaining the tag root for several tokens could significantly complicate the processing of the analyzers results for the purpose of extracting ontology concepts. In addition, MaltParser requires pre-processing of text, unlike UDPipe. In connection with the above and the ease of use of the UDPipe tool itself, the latter was chosen. For further work, we used the pre-trained Russian language model russian-syntagrus-ud-2.4-190531.udpipe, corresponding to the requirements of the Universal Dependencies (UD) framework.

The idea of the Universal Dependencies Framework is to create a cross-language notation for marking up language corpora taking into account the specifics of national languages. The parts-of-speech list is fixed and consists of 17 tags of categories which are subdivided into significant parts of speech, service parts of speech and additional tags. The national language may not support all parts-of-speech tags, but it cannot use additional parts-of-speech tags. The annotated Russian corpus UD-Russian-SynTagRus uses all 17 part-of-speech tags. Lexical and grammatical features of words are set using tags of morphological features. The set of morphological features tags is not predefined like the set of parts-of-speech tags. For each language, specific morphological features are used. A set of relations for cross-language parsing is used to annotate the syntactic dependencies between words. The list of syntax dependencies is fixed and consists of 37 tags. From the predefined tags list of syntactic dependencies in the annotated corpus of the Russian language UD-Russian-SynTagRus, 32 tags are used. Using the UDPipe tool that supports the Russian language model that meets the requirements of the Universal Dependencies project, it is possible to develop a set of concept extraction rules that will be closer to existing English-language analogs of rules using syntactic dependencies. At the same time, the distinctive features of the Russian language are concentrated in a more limited set of relations associated with the annotation of morphological features. It also seems reasonable to reflect the distinctive features of the Russian-language model in the ontology in the form of data properties for words instances that make up domain concepts.

4 An Approach to Extracting Ontology Concepts from the Results of Automatic Processing of Textual Requirements

An ontology system that includes an ontology containing knowledge about requirements artifacts and the requirements engineering process based on the Scrum framework, a application domain ontology and an requirements ontology for analyzing the consistency of a requirements set, we have developed to provide information support for the requirements engineering process with a agile approach [18,19]. In this section, we will consider some classes of the application domain ontology that can be updated based on the proposed approach to extracting concepts, as well as some classes of requirements ontologies that are proposed to be filled using the developed approach.

The life cycle of the proposed ontology is iterative, the development of the project's domain ontology is supposed to be carried out at each iteration of the project as new knowledge about the project appears. In each cycle, the current version of the ontology moves through the development stages: specification, conceptualization, formalization, implementation and maintenance. The effort involved in these activities is not the same throughout the life cycle of an ontology.

The description of the domain area terminology includes the domain concepts and the words that make up the domain concepts. The domain area terminology is proposed to be presented taking into account some data properties, object properties and classes characteristic of lexical ontologies. This will allow us to further compare the data of the ontology of the subject area with the data of external lexical ontologies in the analysis of the data.

First, the basic structure of the ontology was defined, in which the main classes, data and objects properties were identified. Next, logical axioms are defined. The revision of the structure is supposed to be carried out as needed in case of detection of changes in the structure of external sources, on the basis of knowledge about which the structure of the ontology has been developed, as well as when identifying previously undefined logical relationships between concepts. The main development of the ontology is supposed to be carried out by replenishing it with classes instances from requirements texts and lexical relations between words that make up concepts and the actual concepts themselves.

The words that make up the concepts include nouns, verbs, auxiliary verbs, and definitions compatible with nouns. The ontological graph of classes *Domain-Concept* and *Word* is shown in Fig. 5. In the class *DomainConcept*, there are two main subclasses that are needed to describe of requirements: *Entity* and *RelationAction*. Entities, in turn, are divided into actors and objects. An actor is an entity that has some behavior. The role of an actor can be a user or a software

system (subsystem). An object is an entity on which actions performed by an
actor are directed. The actor and the object are interconnected by action relations. The *RelationAction*, in turn, should include a semantic verb, and may also
include an auxiliary (modal) verb and the particle "not". The semantic verbs of
the domain area will be allocated in a separate class *OperationAction*.

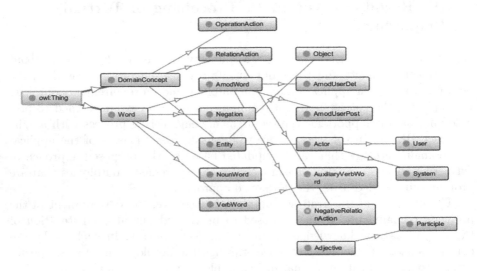

Fig. 5. Ontological graph of classes *DomainConcept* and *Word*.

The requirements ontology is intended to analyze a set of requirements presented in the form of an ontology concepts. The process of extracting ontology concepts from the requirements for subsequent analysis is extremely time-
consuming and requires automation. It should be noted that the relations
between the extracted concepts can be established on the basis of data from the
application domain ontology, where the concepts known to us and the relations
between the concepts or their components verified by the domain area specialist
are stored. Previously unknown concepts can be extracted into the requirements
ontology from text requirements, then external WordNet-like resources can be
used to identify relations between the concepts. Classes, object and data properties for requirements ontology are shown in Fig. 6.

A set of production rules that enables to extract instances of the classes *Actor*,
Action, *Object*, *Word* was developed to process the results of a parsing textual
requirements. For instances of the class *Word*, data properties were defined for
writing tag values of speech parts and morphological features of speech parts in
accordance with the UD-Russian-SynTagRus corpus model. These data properties are used when it is necessary to compare words and the concepts consisting
of them with the data of the application domain ontology, as well as when replenishing the application domain ontology with new concepts.

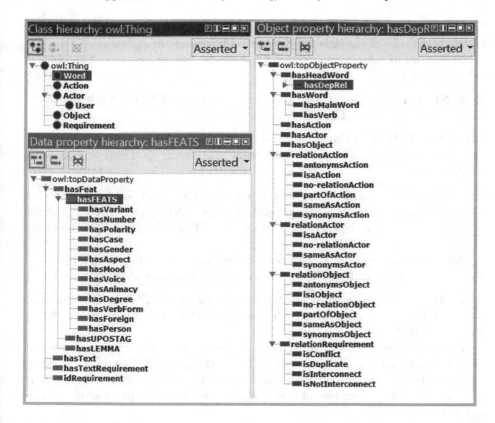

Fig. 6. Classes, object and data properties for Requirements ontology.

Data properties intended for storing the values of parts-of-speech tags and morphological features are grouped as subproperties of the data property *has-FEATS*. The range of accepted values for each of these data properties is limited to the set of valid values from the language model. For example, the data property *hasAnimacy* corresponds to the animate attribute and can take one of three values: Anim, Inan, EMPTY.

Many developed rules can be divided according to purpose into two subsets: the rules for finding the main word in the concept of domain area and the rules for identifying all the words included in the concept. In the beginning, rules are used to define the main words in concepts. These include:

Rule No. 1. Definition of the main word in the entity name: **If** X is a noun **AND** X is connected with the verb Z with the dependency relation $R_{XZ} \in$ {nsubj, obj, parataxis, advcl} **AND** Z is the root of the dependency tree **Then** X is the main word in the name of the entity.

Rule No. 2. Definition of operation-action (semantic verb for relation-action): **If** Y is a verb **AND** Y has a morphological marker VerbForm $\in$ {Inf, Fin} **AND**

Y is not a modal verb **AND** (Y is connected to the root of the dependency tree **OR** Y is the root of the dependency tree) **Then** Y is the operation-action.

Next, a set of rules is applied to determine the full entity name. Examples of rules are given below.

Rule No. 1: If X is the main word in the entity name **AND** X is not associated with any defining word **Then** X is the entity name.

Rule No. 2: If X is the main word in the entity name entity **AND** X is connected with the adjective Z dependency relation R_{XZ} **Then** Z ‖ X is the entity name.

Rule No. 3: If X is the main word in the entity name **AND** X is connected with the noun Z by the dependency relation R_{XZ} **AND** Z is connected with the preposition G by the dependency relation R_{XG} **Then** X ‖ G ‖ Z is the entity name.

After determining the boundaries of the entity, the entity type (actor or object) is determined. Below is one of the rules for finding an object.

Rule: If X is the main word in the entity name S **AND** X is the inanimate noun **AND** X is connected with the predicate P by the dependency relation $R_{PO} \in \{obj\}$ **Then** S is the object.

A group of rules is also used to find the boundaries of the relation-action and identify relation-actions with negation. From the point of view of dependency analysis, the process of extracting the boundaries of the entity name is more complicated, since a compound entity name may, for example, include several adjectives.

When developing the rules, the dependency trees obtained for the sets of requirements were first analyzed. Their analysis showed that out of 32 tags of syntactic dependencies, 7 have never been encountered: csubj and csubj: pass, dep, discourse, expl, iobj, orphan, vocative. Also from the group flat, flat: foreign and flat: name, only the syntactic dependency flat: foreign was encountered. At the moment, the rule database includes 51 rules. The rules set is not exhaustive. It is replenished as previously unknown combinations of syntactic dependencies in combination with morphological features are discovered.

An approach to extracting requirements ontology concepts from the results of automatic processing of textual requirements is as follows:

Step 1. *Processing textual requirements using the Udpipe tool*
A simple text file containing a set of requirements is passed for processing to the Udpipe analyzer.

Result: dependency trees in a CoNLL-U format file.

Step 2. *Extracting the actor, relation-action and object*
The developed extraction rules apply to the results of the previous stage.

Result: a list of sentences indicating for each sentence the extracted names of entities and relation-actions.

Step 3. *Writing data about instances in the requirements ontology*
List items from the previous step are bypass. Instances of classes are added to the ontology, their data properties are indicated, object properties relations between words for the concept are set.

Result: an ontology filled with requirements instances.

To evaluate the performance of the approach, experiments were carried out on a real set of user stories. The set contained 37 user stories. When extracted, the actor and the action were unmistakably highlighted. Errors occurred while extracting objects. 26 concept objects were correctly identified. Let's look at an example of an erroneous extraction. In the user story ".... I want to set the minimum-necessary and planned indicators of the food-hundred ..." when extracting, the "necessary planned indicators of the food-hundred" were obtained. The latter is due to the not very successful formulation of the object itself and, accordingly, the absence of a rule in the base for processing this case. From the point of view of the subject area there are "minimum indicators" and "target indicators". Correctly extracted concepts are validated by the user, after which they can be used to analyze the requirements for the project. As the number of requirements for a project grows, the usefulness of the ontology increases. The updated set of requirements can be checked semi-automatically using previously proven knowledge of domain concepts. This is extremely important for the owner of the product and the development team in the revision of requirements in each iteration of the project, since it helps to identify the appearance of previously unused concepts, which can be either new terms or synonyms of already used ones. In the latter case, one of the synonymous terms should be chosen.

5   Conclusion and Future Work

Domain knowledge is accumulated in parallel with the development of the requirements artifacts. In this paper, an approach to the automatic extraction of requirements ontology concepts was proposed. The structure of the model for representing concepts in the requirements ontology is organized so that it is possible to compare the extracted concepts with already known concepts from the ontology of the application domain. The application domain model of the software product includes the domain terminology which is presented in the form of semantic concepts of the domain area. Each concept may consist of one or more words. A distinctive feature of the domain model is its structural compatibility in terms of representing words with WordNet-like resources which in the future provides the possibility of using the domain ontology together with external resources to identify lexical relations between the words that make up the concepts and then the relations between the concepts.

The proposed approach to extracting requirements ontology concepts from the results of automatic processing of requirements texts is based on the representation of knowledge about syntactic dependency tags, speech parts tags and morphological features of speech parts in the annotated Russian-language UD-Russian-SynTagRus corpus. This text corpus is used by the UDPipe tool. The developed approach can be applied to other models of the Russian language that correspond to syntactic relations for Universal Dependencies. The next stage of our research will be the development of the software that will allow its user to quickly create and edit rules in design mode. As well as adding rules for searching for wording of requirements with indicating several objects to display recommendations for improving the wording of the requirements.

Acknowledgments. The research is supported by Ministry of Science and Higher Education of Russian Federation (project No. FSUN-2020-0009).

References

1. Alsanad, A.A., Chikh, A., Mirza, A.: A domain ontology for software requirements change management in global software development environment. IEEE Access **7**, 49352–49361 (2019). https://doi.org/10.1109/ACCESS.2019.2909839
2. Kibret, N., Edmonson, W., Gebreyohannes, S.: Ontology-driven requirements engineering in the responsive and formal design process. In: Adams, S., Beling, P.A., Lambert, J.H., Scherer, W.T., Fleming, C.H. (eds.) Systems Engineering in Context, pp. 395–405. Springer, Cham (2019). https://doi.org/10.1007/978-3-030-00114-8_33
3. Dermeval, D., Vilela, J., Bittencourt, I.I., et al.: Applications of ontologies in requirements engineering: a systematic review of the literature. Requir. Eng. **21**, 405–437 (2016). https://doi.org/10.1007/s0076601502226
4. Kaiya, H., Saeki M.: Ontology based requirements analysis: lightweight semantic processing approach. In: Fifth International Conference on Quality Soft-ware (QSIC 2005), Melbourne, Victoria, Australia, pp. 223–230. IEEE (2005). https://doi.org/10.1109/QSIC.2005.46
5. Jaiwai, M., Sammapun, U.: Extracting UML class diagrams from software requirements in Thai using NLP. In: 14th International Joint Conference on Computer Science and Software Engineering (JCSSE), pp. 1–5. IEEE, Nakhon Si Thammarat (2017). https://doi.org/10.1109/JCSSE.2017.8025938
6. Herchi, H., Abdessalem, W.B.: From user requirements to UML class diagram. In: Proceedings of the International Conference on Computer Related Knowledge, Sousse, Tunisia, pp. 19–23 (2012)
7. Arellano, A., Zontek-Carney, E., Austin, M.A.: Natural Language Processing of Textual Requirements. In: The Tenth International Conference on Systems (ICONS 2015), Barcelona, Spain, pp. 93–97 (2015)
8. Arellano, A., Zontek-Carney, E., Austin, M.A.: Frameworks for natural language processing of texual requirements. Int. J. Adv. Syst. Measur. **8**(3&4), 230–240 (2015)
9. Schlutter, A., Vogelsang, A.: Knowledge representation of requirements documents using natural language processing. In: REFSQ 2018 Joint Proceedings of the Co-Located Events NPL4RE: 1st Workshop on Natural Language Processing for Requirements Engineering, vol. 2075. RWTH, Aachen (2018). https://doi.org/10.14279/depositonce-7776

10. Robeer, M., Lucassen, G., van Der Werf, J.M.E.M., Dalpiaz, F., Brinkkemper, S.: Automated extraction of conceptual models from user stories via NLP. In: 24th International Requirements Engineering (RE) Conference, pp. 196–205 (2016)
11. Yarushkina, N., Filippov, A., Moshkin, V., Egorov, Y.: Building a domain ontology in the process of linguistic analysis of text resources. Yarushkina, Preprints 2018, pp. 1–20 (2018). https://doi.org/10.20944/preprints201802.0001.v1
12. Assawamekin, N., Sunetnanta, T., Pluempitiwiriyawej, C.: Ontology-based multi-perspective requirements traceability framework. Knowl. Inf. Syst. **25**(3), 493–522 (2010). https://doi.org/10.1007/s10115-009-0259-2
13. Boukhari, I., Bellatreche, L., Khouri, S.: Efficient, unified, and intelligent user requirement collection and analysis in global enterprises. In: Proceedings of International Conference on Information Integration and Web-based Applications and Services (IIWAS 13), pp. 686–691. Association for Computing Machinery, New York (2013). https://doi.org/10.1145/2539150.2539263
14. Khouri, S., Boukhari, I., Bellatreche, L., Jean, S., Sardet, E., Baron, M.: Ontology-based structured web data warehouses for sustainable interoperability: requirement modeling, design methodology and tool. Comput. Ind. **63**(8), 799–812 (2012). https://doi.org/10.1016/j.compind.2012.08.001
15. Bures, T., Hnetynka, P., Kroha, P., Simko, V.: Requirement specifications using natural languages. Charles University Faculty of Mathematics and Physics. Dep. of Distributed and Dependable Systems. Technical Report D3S-TR-2012-05 (2012)
16. Sharoff, S.: Russian statistical taggers and parsers. http://corpus.leeds.ac.uk/mocky/. Accessed 5 May 2020
17. Sharoff, S., Nivre, J.: The proper place of men and machines in language technology. In: Processing Russian Without any Linguistic Knowledge: Proceedings of Dialogue 2011, Russian Conference on Computational Linguistics (2011)
18. Murtazina, M., Avdeenko, T.: An ontology-based approach to the agile requirements engineering. In: Bjørner, N., Virbitskaite, I., Voronkov, A. (eds.) PSI 2019. LNCS, vol. 11964, pp. 205–213. Springer, Cham (2019). https://doi.org/10.1007/978-3-030-37487-7_17
19. Murtazina, MSh., Avdeenko, T.V.: Requirements analysis driven by ontological and production models. In: CEUR Workshop Proceedings, vol. 2500, pp. 1–10 (2019)

Data Driven Detection of Technological Trajectories

Sergey Volkov[1,2]($\boxtimes$), Dmitry Devyatkin[1], Ilya Tikhomirov[1], and Ilya Sochenkov[1]

[1] Federal Research Center "Computer Science and Control" RAS, Moscow, Russia
[2] Peoples' Friendship University of Russia (RUDN University), Moscow, Russia

Abstract. The paper presents a text mining approach to identifying and analyzing technological trajectories. The main problem addressed is the selection of documents related to a particular technology. These documents are needed to detect a trajectory of technology. The approach includes new keyword and keyphrase detection method, word2vec embeddings-based similar document search method and fuzzy logic-based methodology for revealing technology dynamics. USPTO patent database was used for experiments. The database contains more than 4.7 million documents from 1996 to 2020. Self-driving car technology was chosen as an example. The result of the experiment shows that the developed methods are useful for effective searching and analyzing information about given technologies.

Keywords: Technological trajectories · Text mining · Keywords · Similar documents retrieval

1   Introduction

Giovanni Dosi defines "technological paradigm" as a specific set of technological innovations; which has its list of "relevant" problems as well as the procedures and the specific knowledge to solve these problems [1]. The development path within a technological paradigm is a technological trajectory.

Although those concepts are well-established, there is still a lack of data-driven methods which could reveal that trajectories in large technology-related corpora, such as patent or research article databases. The main obstacles to the development of such an approach are the following.

1. There is a need for methods and tools for technology-related information retrieval. Existing tools rely mainly on keyword-based search and topic similarity. However, the topic similarity between patents does not always reflect the closeness of the claimed technologies.
2. There is a lack of formal models which can predict changing of a technological trajectory.

© Springer Nature Switzerland AG 2021
A. Sychev et al. (Eds.): DAMDID/RCDL 2020, CCIS 1427, pp. 204–215, 2021.
https://doi.org/10.1007/978-3-030-81200-3_15

In the paper, we propose a text mining approach to identifying and analyzing technological trajectories. To deal with the first issue, we propose a new technology keyword and keyphrase detection method, which combines topic importance characteristic and neology score of terms. Then we utilize word embeddings-based method for identification of a technological trajectory [2] and apply the new keyword score to make the search more technology-focused. In the analyzing part, we mainly consider the conditions which precede a breakthrough. To overcome the second issue, we created a fuzzy logic-based method for revealing technology dynamics. The proposed approach contains several essential steps:

1. Selecting the information sources, crawling and indexing documents.
2. Document retrieval from the index database and selecting the subset of documents which are related to the technology, which is the point of interest.
3. Automated analysis of metadata on the previously selected documents. Experts study publication and patent activity, primary patent right holders and R&D centers; then this information is used to identify how the technology has been developing.

Finally, we used the USPTO patent database to verify the proposed approach and chose self-driving car technology as a sample for the experiments.

2 Related Work

The primary researches are focused on identifying the conditions which precede sky-rocketing of a technology. Namely, [3] proposes the theory of technological parasitism that is based on the idea of parasite-host relationships between technologies. Technologies with a high number of technological parasites have an accelerated evolution induced with symbiosis-like relationships. Those symbioses provide the basis for extensive development and adaptive behavior of interactive technologies in markets. The importance of interaction between technologies is also highlighted in the researches by Loreto and Napolitano with colleagues [4, 5]. They claim that the context of innovation (like related or close technologies) defines the development trajectory of the innovation. The experiments on a statistic model show the correctness of that claim.

Apparently the generally recognized idea is that the most prospective technologies are the new ones, which have a good support in terms of research, development and marketing [6]. However, there is evidence, which show alternative pathways of technological development. For example, Ardito [7] study a phenomenon of established technologies, which anyway have become a breakthrough. They have checked two hypotheses. The first one is that the probability of technology growth increases with its level of establishment and then decreases, thus having an inverted U-shaped form. The second one is that the size of knowledge base of established technologies decreases the positive linear dependence between established technologies and the probability of becoming breakthroughs. They checked the both claims statistically with USPTO database and have confirmed them.

Another bunch of researches is devoted to trajectory identification. Zhou et al. [8] propose a framework for the identification of emerging technologies. The framework includes Newmann topology clustering [9] for convergence cluster detection, LDA topic modeling for cluster topics identification, and network visualization with the Citation-Network Data Analyzer (CDA) for interdisciplinary citation identification. They form the technological trajectory convergence with the transition patterns of clusters in adjacent periods. The clusters whose overlap reaches a predefined threshold are considered to have a transition relationship, the clusters with a transition relationship are linked, and topics and times are labeled. In the paper [10] researchers propose a method for assessment of patent similarity with semantic text analysis. They show the main advantages of the method in comparison to others, such as (IPC-code analysis, citation analysis or keyword analysis). Text similarity approach is also used in [11], where they seek early patents, similar to the analyzed one.

Norman with colleagues [12] propose an original feature which allows revealing technology-related keywords in large datasets. They conducted experimental research, which shows that a frequency of using technical-related terms in documents is distributed non-uniformly, while a frequency of common lexis holds stable. This feature appears to be important not for keyword generation only, but also for searching for the technology-related documents.

The review shows that the context of innovation holds crucial information for revealing technological trajectories. Therefore we propose to extract and estimate the context of the analyzed technology, but also to utilize a fuzzy logic framework for that estimation [13] to make the results human-interpretable. In order to obtain accurate technological trajectory identification, we decided to adopt the novelty scoring from [12] for searching technology-related patents.

3 Dataset

The United States Patent and Trademark Office (USPTO) database is the primary data source for the experiments [14]. The database contains more than 4.7 million documents from 1996 to 2020 years. Basically, it is a set of web pages; each of them corresponds to one patent. The database has been treated in the following way. First, we have applied Text Appliance web crawling tools [15] to download those web-pages, and then chose the technology for the test, "Self-driving car". After that, we manually selected 10 "seed" patents with the integrated USPTO search system.

Then we assembled a gold dataset to evaluate the accuracy of the presented methods with the following steps. First, we built a list that combines search results of all the considered methods with the results obtained by human analytics with the USPTO search tool. All 4.7 million documents were used to perform the searches. Second, we submitted that list of documents to human annotators who excluded all non-relevant documents from the list. Finally, all the remaining patents are related to the self-driving car technology or the technologies that could have preceded it. Thus, when calculating the accuracy, the found patent is considered valid if the gold dataset contains it [16].

4 Description of Methods

4.1 Method for an Identification of a Technology

The method contains the following steps. The first step is keyword and key-phrase extraction for the seed documents. In this case keywords are the most important lexis of a document [17]. We applied well-known *tf-idf* score to assess that importance:

$$tf(t, d) = \frac{n_t}{\sum_k n_k}$$

there n_t – count of the token t in the document d. Denominator is a total number of tokens in the document d.

$$idf(t, D) = \log \frac{|D|}{|\{d_i \in D | t \in d_i\}|}$$

$|D|$ - count of documents in the analyzed collection.
$|\{d_i \in D | t \in d_i\}|$ - count of documents (in D), which contain t.

$$tf\text{-}idf(t, d, D) = tf(t, d) \times idf(t, D) \tag{1}$$

Similarly to the formula (1) for a lexical descriptor w in a text document t and topic class c, we consider the quantity

$$TFtIDF(w, c, t) = TF(w, t) \cdot \Delta I^+(w, c)$$

which is called topic importance characteristic [17]. It can be used for getting keywords of a whole collection of documents. All obtained keywords should be stored separately as well as their weights.
After that the document dates are used to re-weight the keywords. For that we utilize a novelty score which is a frequency of using of some word in a particular date interval.

$$nw = tf\text{-}idf(t, d, D) * c$$

$$c = \frac{avg(P_{ij})}{avg(P) * D_{count}(i, j)}$$

P_{ij} – count of the word occurrences from the year i to year j.
D_{count} – documents count from $(i - j)$ interval.
In this study we consider 5 date intervals from 1996 to 2020; therefore we obtain particular value for an each interval.

Third step is sorting keyword and keyphrases with the obtained weights. We obtain top-50 keywords for each pair *<interval, document>* and use them in the following steps.

Forth step includes training Word2Vec models on documents, related to the analyzed intervals (2013–2020, 2009–2014, 2005–2010, 2000–2006, 1996–2001). We got 5 different models as the result.

The next step is seeking documents that are similar to the previously obtained ones.

1. Vectorize all keywords from the current document list with the model for the current interval (begin with «2013–2020»).
2. Sum the obtained vectors for each document and form the result vector. This way each document is represented with one vector.
3. Search for similar documents in the whole patent collection. We use cosine similarity for evaluating the distance between documents.
4. Extend the current document list by choosing top-n of the most similar document from the obtained in step 3.
5. Repeat all steps for the earlier date interval.

4.2 Method for Revealing of Technology Dynamics

Consider per-year cumulative patent count for the identified technologies. We applied logistic curve (S-curve) [18] for the approximation of the cumulative patent count f : $\mathbb{R} \to \mathbb{R}^+, f(x) = \frac{C}{b+e^{-ax}}$, where $a, b, C \epsilon \mathbb{R}^+$, – parameters of that function are obtained with the least-squares approach, and x is a normed year (Fig. 1). The applicability of that curve for technology growth modeling has been proved, for example in [19].

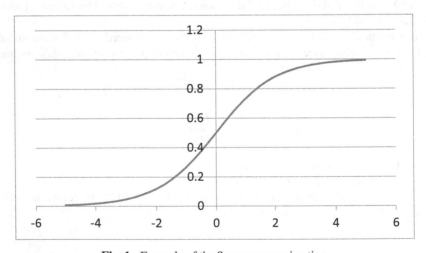

Fig. 1. Example of the S-curve approximation.

The cumulative patent count has the highest speed of growth if the first derivative of the S-curve function reaches the maximum:

$$f'(x) = \frac{Cae^{-ax}}{(b + e^{-ax})^2}$$

Let find the argument x_{max}, for that maximum:

$$f''(x) = -\frac{Ca^2\left(1 - \frac{2e^{-ax}}{b+e^{-ax}}\right)e^{-ax}}{(b + e^{-ax})^2}$$

$$f''(x_{max}) = 0, x_{max} = -\frac{ln(b)}{a}$$

Define a linguistic variable "Technology development dynamics", which can take values from the set $X = \{X_h, X_l, X_m\}$, there $X_h = $ 'high', $X_l = $ 'low', $X_m = $ 'middle'.

Define a fuzzy set for each value of the linguistic variable. Those fuzzy sets contain arguments of $f(x)$. Let define membership functions for that:

$$\mu_{h,m,l}: \mathbb{R} \rightarrow [0, 1]$$

First of all, define a membership function for the set which is related to X_h. As said earlier speed of patent growth depends on $f'(x)$. That derivative is always positive and upper-bounded. Accordingly to the condition $\mu \leq 1$ we find it uniform norm:

$$f'_{max} = \frac{Ca}{4b}$$

$$\mu_h(x) = \frac{f'(x)}{f'_{max}}$$

Hence value "low" is semantically negative to "high", one can define the membership function for that value as follows:

$$\mu_l(x) = 1 - \mu_h(x)$$

Value "middle" describes situation, there $\mu_l(x) = \mu_h(x)$, therefore the membership is the following (Fig. 2):

$$\mu_m = 1 - |\mu_h(x) - \mu_l(x)|$$

Finally, if one wants to evaluate the development dynamics of technology at some year x, they should choose the membership function which returns the maximum and then find the value of the linguistic variable for that function. Accordingly to [5], we assess the state of the analyzed technology as well as states for the related technology on particular dates and try to find a relationship between them.

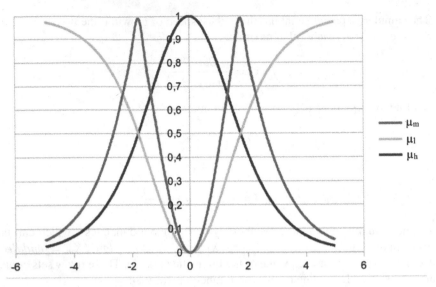

Fig. 2. Examples for the membership functions μ_h, μ_m and μ_l

5   Results

Our assumption is based on the theory that every technological term has a different usage popularity at different time periods. Therefore, if we take into account the popularity of the term in a certain period; we can get a batch of more relevant keywords. Because of that, technological terms will have a higher coefficient than common words, even if there are more common words in a sentence.

Figure 3 shows that words matching some terms have a non-uniform distribution unlike common words. For example, such words as "method", "right", "light" have smooth horizontal curves while words "car", "autonomous", "pen" have curves with pronounced peaks. Given this feature, we can find more relevant document which consists information about technology. Next table shows result of sorting keywords for seed selection of documents (Table 1).

Similar documents retrieval started with that sorting keywords method. Two methods were compared. First method is sequential search. There are five steps corresponding to five time periods. On each step similar documents retrieval applies only to the documents, which were found on the previous step. Four cases were considered (for top 5, 10, 20, 50 keywords which are sorted by new coefficient). Next table shows step-by-step accuracy. On each step 50 documents were found. The document is considered valid if it is contained in the experts-filtered gold dataset. We used different word2vec models to find documents on each step (models were trained on documents from different time interval). We also used the novelty coefficients corresponding to a given time interval at each stage for all documents (Table 2).

The second method is the full search. It differs from the first one in that similar documents retrieval is processed at each step not only for patents, found on the previous

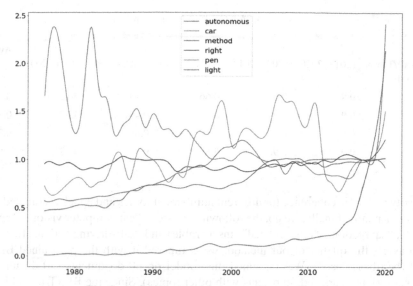

Fig. 3. Example of novelty scores for some words

Table 1. NDCG of keywords batch relevance

Top-N\method	TF-IDF+proposed (2013–2020)	TF-IDF
5	0.915	**0.926**
10	**0.944**	0.920
20	**0.866**	0.826
50	**0.838**	0.805

Table 2. Accuracy of sequential search

Step (interval)/Top-N Keywords	1 (2013–2020)	2 (2009–2014)	3 (2005–2010)	4 (2000–2006)	5 (1996–2001)	Avg
5	**0.9**	0.66	0.68	0.7	0.26	0.64
10	**0.9**	0.2	0.16	0.62	0.02	0.38
20	0.88	0.46	0.56	0.86	**0.96**	0.74
50	**0.9**	**0.96**	**0.8**	**0.94**	0.92	**0.904**

step, but for all current documents. Next table shows step-by-step accuracy. On each step, about 50 documents were found (Table 3).

After that, we clustered the all obtained document, so as to separate self-driving car technology patents from the ones related to auxiliary technologies. Latent Dirichlet allocation (LDA) model [20] was used to solve the clustering task. Having levels of

Table 3. Accuracy of full search

Step (interval)/Top-N Keywords	1 (2013–2020)	2 (2009–2014)	3 (2005–2010)	4 (2000–2006)	5 (1996–2001)	Avg
5	**0.9**	0.66	0.66	0.14	0.12	0.5
10	**0.9**	0.2	0.32	0.48	0.02	0.38
20	0.88	0.46	0.56	0.86	1	0.75
50	**0.9**	**0.96**	**0.8**	**0.9**	0.98	**0.908**

perplexity and topic coherence for different numbers of topics and epochs estimated, we set them manually. Finally, we got the following clusters: "car computer vision", "traffic control", "navigation", "other related" (inseparable) and "self-driving car" itself.

To assess the quality of our method, we compared it with the pre-trained BERT model (bert-base-cased). We fine-tuned this model on 60 documents (30 documents about self-driving cars and 30 patents with other topics). Since the BERT model has a limitation on the length of the texts, which is 512 tokens, we used only abstracts of the patents and split them into fragments. More precisely, the dataset we used contains pairs of text fragments, and each pair consists of two fragments from different documents. The model was trained to predict if those pairs are related to the same topic. Thus, more than 8K pairs were used for the training.

After that, the process of searching for patent began using the BERT model. The process is divided into several steps. The search starts with the same 10 seed documents. On each next step, the algorithm searches similar patents for the previously found documents. Table 4 shows step-by-step accuracy. Although BERT performs slightly better in the initial steps, it shows lower accuracy in total. We believe this is because the novelty score helped to filter non-technical lexis, which allows the proposed method to stay on the track.

Table 4. Comparison of the proposed method with the pre-trained BERT model.

	1	2	3	4	5	6
BERT	**0.92**	0.821	**0.824**	0.879	0.865	0.8
Proposed	0.9	**0.96**	0.8	**0.9**	**0.98**	**0.908**

Finally, we analyzed the cumulative patent dynamics for the self-driving-related technologies and built fuzzy-set membership functions for them. The μ_h functions which determine the high level of technology development dynamics are presented in Fig. 4. The graph illustrates the idea that a breakthrough is preceded with high development dynamics of the related technologies, namely the highest dynamics for "car computer vision", "traffic control", "navigation" and other technologies was in 2012–2017. In contrast, the dynamics for self-driving cars (denoted as "autonomous driving") was still relatively low.

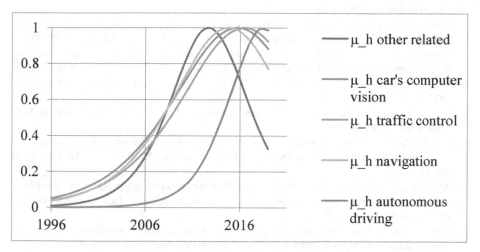

Fig. 4. $\mu_h(x)$ for the analyzed technologies

6 Conclusion

The result of the experiment shows that the developed methods are useful for technology-related document retrieval and breakthrough-technologies prediction. Unfortunately, it is not tricky to detect a technological trajectory using a single source. In the future, we are going to increase the number of information sources and collect extensive statistics on the development of various technologies. Namely, we are going to consider not only patents but also scientific papers, theses, technical reports, and other sources.

The future development of the proposed method would also be combining the keyword weight feature with sentence-level embedding models, such as Doc2Vec [21], or with language models [22]. This way, one could consider a context of the technology-related lexis and achieve better accuracy of the retrieval.

The detailed analysis of obtained patents shows how the technologies develop in time and predict possible future directions of that development. In addition to a breakthrough prediction, we are going to consider additional technology-state related events, such as the emergence of new technology.

Acknowledgements. This study was supported by Russian Foundation for Basic Research, grant number 17-29-07016 of i_m.

References

1. Dosi, G.: Technological paradigms and technological trajectories: a suggested interpretation of the determinants and directions of technical change. Res. Policy **11**(3), 147–162 (1982)

2. Volkov, S.S., Devyatkin, D.A., Sochenkov, I.V., Tikhomirov, I.A., Toganova, N.V.: Towards automated identification of technological trajectories. In: Kuznetsov, S.O., Panov, A.I. (eds.) Artificial Intelligence: RCAI 2019, pp. 143–153. Springer, Cham (2019). https://doi.org/10.1007/978-3-030-30763-9_12

3. Coccia, M., Watts, J.: A theory of the evolution of technology: technological parasitism and the implications for innovation management. J. Eng. Technol. Manag. **55**, 101552 (2020)

4. Loreto, V., Servedio, V.D.P., Strogatz, S.H., Tria, F.: Dynamics on expanding spaces: modeling the emergence of novelties. In: Degli Esposti, M., Altmann, E.G., Pachet, F. (eds.) Creativity and Universality in Language. LNM, pp. 59–83. Springer, Cham (2016). https://doi.org/10.1007/978-3-319-24403-7_5

5. Napolitano, L., et al.: Technology networks: the autocatalytic origins of innovation. R. Soc. Open Sci. **5**(6), 172445 (2018)

6. Luo, C., Zhou, L., Wei, Q.: Identification of research fronts in artificial intelligence. In: 2017 2nd Asia-Pacific Conference on Intelligent Robot Systems (ACIRS), pp. 104–108. IEEE (2017)

7. Ardito, L., Messeni, P.A., Panniello, U.: Unveiling the breakthrough potential of established technologies: an empirical investigation in the aerospace industry. Technol. Anal. Strateg. Manag. **28**(8), 916–934 (2016)

8. Zhou, Y., et al.: Unfolding the convergence process of scientific knowledge for the early identification of emerging technologies. Technol. Forecast. Soc. Chang. **144**, 205–220 (2019)

9. Zou, W., et al.: Clustering approach based on Von Neumann topology artificial bee colony algorithm. In: Proceedings of the International Conference on Data Mining (DMIN) – The Steering Committee of The World Congress in Computer Science, Computer Engineering and Applied Computing (WorldComp), p. 1 (2011)

10. Wang, X., Ren, H., Chen, Y., Liu, Y., Qiao, Y., Huang, Y.: Measuring patent similarity with SAO semantic analysis. Scientometrics **121**(1), 1–23 (2019). https://doi.org/10.1007/s11192-019-03191-z

11. Kelly, B., et al.: Measuring technological innovation over the long run. National Bureau of Economic Research. No. w25266 (2018)

12. Norman, C.: Technical Term Extraction Using Measures of Neology (2016)

13. Zadeh, L.A.: The concept of a linguistic variable and its application to approximate reasoning—I. Inf. Sci. **8**(3), 199–249 (1975). https://doi.org/10.1016/0020-0255(75)90036-5

14. Search for patents–USPTO. https://www.uspto.gov/patents-application-process/search-patents. Accessed 30 July 2020

15. Osipov, G., Smirnov, I., Tikhomirov, I., Sochenkov, I., Shelmanov, A.: Exactus expert—search and analytical engine for research and development support. In: Hadjiski, M., Kasabov, N., Filev, D., Jotsov, V. (eds.) Novel Applications of Intelligent Systems, pp. 269–285. Springer, Cham (2016). https://doi.org/10.1007/978-3-319-14194-7_14

16. Dataset trajectories-uspto-v2. http://nlp.isa.ru/trajectories-uspto-v2. Accessed 19 May 2020

17. Suvorov, R.E., Sochenkov, I.V.: Establishing the similarity of scientific and technical documents based on thematic significance. Sci. Tech. Inf. Process. **42**(5), 321–327 (2015). https://doi.org/10.3103/S0147688215050081

18. Foster, R.N.: Working the S-curve: assessing technological threats. Res. Manag. **29**(4), 17–20 (1986)

19. Andersen, B.: The hunt for S-shaped growth paths in technological innovation: a patent study. J. Evol. Econ. **9**(4), 487–526 (1999)

20. Blei, D.M., Ng, A.Y., Jordan, M.I.: Latent Dirichlet allocation. J. Mach. Learn. Res. **3**(Jan), 993–1022 (2003)
21. Le, Q., Mikolov, T.: Distributed representations of sentences and documents. In: International Conference on Machine Learning, pp. 1188–1196 (2014)
22. Peters, M.E., et al.: Deep contextualized word representations. In: Proceedings of NAACL-HLT, pp. 2227–2237 (2018)

Comparison of Cross-Lingual Similar Documents Retrieval Methods

Denis Zubarev[(✉)] [iD] and Ilya Sochenkov [iD]

Federal Research Center 'Computer Science and Control' of Russian Academy
of Sciences, 44-2 Vavilov Street, Moscow 119333, Russia
zubarev@isa.ru

Abstract. In this paper, we compare different methods for cross-lingual similar document retrieval for distant language pair, namely Russian and English languages. We compare various methods among them: classical Cross-Lingual Explicit Semantic Analysis (CL-ESA), machine translation methods and approaches based on cross-lingual embeddings. We introduce two datasets for evaluation of this task: Russian-English aligned Wikipedia articles and automatically translated Paraplag. Conducted experiments show that an approach with inverted index, with an extra step of mapping top keywords from one language to other with the help of cross-lingual word embeddings, achieves better performance in terms of recall and MAP than other methods on both datasets.

Keywords: Cross-lingual document retrieval · Cross-lingual plagiarism detection · Cross-lingual word embeddings

1 Introduction

This paper is a continuation of our previous study [28]. Document retrieval from a big collection of texts is important information retrieval problem. This problem is extensively studied for short queries, such as user queries to search engines. The document retrieval with texts as queries imposes some difficulties, among them an inability to capture the main ideas and topics from the long text. The problem becomes even harder when we enter the field of cross-lingual document retrieval. Some tasks require to use a text (possibly long) as a query to retrieve documents that are somehow similar to it. One of these tasks is plagiarism detection that is divided into two stages: source retrieval and text alignment.

On the *source retrieval stage* for a given suspicious document, we need to find all sources of probable text reuse in a large collection of texts. For this task, a source is a whole text, without details of what parts of this document were plagiarized. Typically, we get a large set of documents (around 500 or more) as a result of this stage.

On the *text alignment* stage we compare the suspicious document to each candidate to detect all reused fragments, and identify its boundaries [8,9,18,29].

In this work, we study only the first task. The same stages are valid for cross-lingual plagiarism detection. Given a query document in one language the goal is to find the most similar documents from the collection in another language.

© Springer Nature Switzerland AG 2021
A. Sychev et al. (Eds.): DAMDID/RCDL 2020, CCIS 1427, pp. 216–229, 2021.
https://doi.org/10.1007/978-3-030-81200-3_16

2 Related Work

Some works were recently devoted to the monolingual document retrieval for long texts. In [12] authors introduce a siamese multi-depth attention-based hierarchical recurrent neural network that learns the long text semantics. They conducted multiple experiments, including retrieval of similar Wikipedia articles. In [11] authors try to employ standard approximate nearest neighbour (ANN) search instead of the usual discrete inverted index, for retrieving documents. They learned similarity function and showed that it could improve performance on two similar-question retrieval tasks. However, using the custom similarity functions makes it impossible to employ existing frameworks for ANN; consequently, they used exact search in experiments. In [27] is introduced a framework for monolingual and cross-lingual information retrieval based on cross-lingual word embeddings. They represent user queries and documents as averaged embeddings of words and employ exact search to find similar documents for a given query. The overview of different approaches for cross-lingual source retrieval is presented in [5] and [16]. Also, there made an evaluation and a detailed comparison of some featured methods. In [4], NMT (neural machine translation) is used to translate a query document to the other language. They solve the source retrieval task by employing shingles (overlapping word N-grams) method. They use word-class shingles, instead of word shingles, where each word is substituted by the label of the class it belongs to. To obtain word classes, they apply agglomerative clustering on word embeddings learned from English Wikipedia. In [14] is described a training of word embeddings on comparable monolingual corpora and learning the optimal linear transformation of vectors from one language to another (there were used Russian and Ukrainian academic texts). Also, there were discussed usage of those embeddings in source retrieval and text alignment subtasks. This work focuses on the comparison of inverted index based approaches with ANN approach for distant language pair.

3 Document Retrieval Methods

In this section, we describe various methods that we used for similar document retrieval.

3.1 Preprocessing

On a preprocessing stage, we split each sentence into tokens, lemmatize tokens and parse texts. We use Udpipe[1] [22] for the Russian language and for the English language. Besides, we removed words with non-important part of speech: conjunction, pronoun, preposition, etc., and common stop-words (be, the, etc.).

---

[1] russian-syntagrus-ud-2.5–191206 and english-ewt-ud-2.5–191206 models.

3.2   Cross-Lingual Embeddings

We train cross-lingual word embeddings for a Russian-English pair on parallel sentences available on the Opus site [24] namely:

- News Commentary
- TED Talks 2013
- MultiUN
- OpenSubtitles
- Wiki
- JW300
- QED
- Tatoeba

We extend this corpus with sentences from the Yandex Parallel corpus[2] [1].

All parallel sentences are preprocessed. After that, all pairs that have a difference in the size of more than 10 words are filtered out. We use syntactic phrases up to 3 words in length to enrich the vocabulary. We take only noun phrases and prepositional phrases, which includes the preposition "of" and that are common for the corpus (>20 occurrences). Each sentence with phrases is complemented with the same sentence but without phrases - only with words. Finally, we assembled a corpus of more than 47 million sentences. The dictionary size was around 1.1 million words/phrases.

We apply a method proposed in [26], designed for learning bilingual word embeddings from a non-parallel document-aligned corpus, but it can be used for learning on parallel sentences too. We assume that the structures of the two sentences are similar. Words are inserted into the pseudo-bilingual sentence relying on the order in which they appear in their monolingual sentences and based on a length ratio of two sentences. For example, if we were given two sentences: "Mama myla ramu" and "Mother washed beautiful frame", the result of their merging is "Mama mother myla washed ramu beautiful frame". Since we removed auxiliary words from sentences, we assume that corresponding Russian and English words are in the same context window. It would not be the case if there were a different word order, so we experimented with different window sizes and chose size == 10. After that, the word2vec skip-gram model [15] is used on the resulting bilingual corpus. We use gensim word2vec implementation with those parameters: dimensionality of embeddings was 384, a window size of 10 words, the minimal corpus frequency of 10, negative sampling with 20 samples, no down-sampling, 15 iterations over the corpus.

Embeddings are post-processed in an unsupervised manner using all-but-the-top method [23], which eliminates the common mean vector and a few dominating directions (according to PCA). We empirically chose hyper-parameter dd_A value (10), which denotes the number of dominating directions to remove from the initial representations.

---

[2] Russian-English parallel corpus: https://translate.yandex.ru/corpus?lang=en/.

We evaluate learned embeddings on Multi-SimLex dataset [25]. In the following Table 1 are Spearman's ρ correlation scores for the embeddings from the paper and ours.

Table 1. Embeddings Spearman's scores on Multi-SimLex dataset

	Spearman	Number of OOV
Vecmap unsupervised	0.511	291
Vecmap supervised 5k + self learning	0.551	291
Ours embeddings	0.552	83

In that paper, the embeddings were obtained using Vecmap framework [2], which learns a transformation matrix that maps representations in one language to the representations of the other language. Pretrained FastText embeddings (CC+Wiki) were used as monolingual embeddings. The absolute scores of Vecmap embeddings and ours are not directly comparable, because these models have different coverage. In particular, our embeddings contains less number of out-of-vocabulary (OOV) words, whose FastText embeddings are not available. Also, we evaluated the retrieval capabilities of learned embeddings on the bilingual dictionary of MUSE project[3] [7]. The top 10 most similar words were retrieved for each Russian word in the dictionary and compared with the actual translations (Table 2).

Table 2. Results of embeddings evaluation on the task of dictionary induction

Recall	0.74
Recall w/o OOV words	0.86
P@1	0.77

The size of the dictionary was 35k pairs; the number of OOV words was 3802. Excluding those words from the evaluation gives a more accurate estimate on the model performance (Recall w/o OOV words in the table).

3.3 Inverted Index Based Approach

We use a custom implementation of inverted index [20], which maps each word to a list of documents in which it appears along with weight (e.g. TF) that represents the strength of association of this word with a document. Along with words, we index syntactic phrases up to 3 words.

---

[3] A library for Multilingual Unsupervised or Supervised word Embeddings https://github.com/facebookresearch/MUSE.

At query time, we extract the top terms from the query document according to some weighting scheme. Then we map each keyword to N other language keywords with cross-lingual embeddings. If a keyword is not in embeddings, then this word is added as is, since it may be the word in the other language. We preserve the weights of keywords from the original top. The searcher retrieves corresponding documents from the inverted index using query keywords and merges them into weighted vectors of keywords that represent the other documents. Then we use either classical ranking function like BM25 or calculate similarity score between the query vector and all other vectors (cosine similarity or hamming similarity). It should be noted that comparison via similarity measures is asymmetrical since vectors of other documents consist only of words from the query vector. Although it is not the most accurate representation of these documents, the comparison is very efficient, and retrieval performance (recall) is not affected much by that. When using similarity functions (cosine or hamming), terms are weighted with TF*IDF, where $TF = \log_{\text{len}(D)+1}(\text{Cnt}(w_i) + 1)$ for word w_i from a document D, and $\max(0, \log_{10}(N - w_{\text{cnt}} + 0.5)/(w_{\text{cnt}} + 0.5))$ is IDF, where N is the total amount of documents in a collection.

3.4   Translation Method

This method resembles a translation method that was introduced in [17]. The fundamental idea of the translation methods is, for each term t of a query q, to replace any existing related terms t' in a document d with the term itself, but counting its occurrence as less than 1. Basically, this method is the same as the previous one, with the exception that related terms are merged into the term, to which they are related, after the retrieval stage. The weight of each term is adjusted according to the formula:

$$\widehat{w}_t = w_t + \sum_{t' \in R(t)} P_T(t|t')w_{t'} \tag{1}$$

where $R(t)$ is a set of related terms, $P_T(t|t')$ is the translation probability. In this case, translation probability is a similarity score, which value is expected to be between 0 and 1.

3.5   Machine Translation

It is quite a natural approach: translate the query text to the other language and perform monolingual retrieval of similar documents. For training machine translation model, we used a subset of the same parallel sentences as for training cross-lingual embeddings, except OpenSubtitles sentences. Also, we removed all sentence pairs with the difference in length greater than 4 words. We used OpenNMT-py library[4] to train a machine translation model: RNN encoder-decoder with attention and mostly default settings. Russian and English dictionary sizes: 181 k and 114 k, respectively.

---

[4] version 1.1.1.

3.6   Document as Vector

In this approach, we represent each document as a dense vector. It is done by averaging vectors of the top K keywords of the document. After that, we index all vectors with ANN index. At query time, the given document is transformed into the vector representation, and the approximate nearest neighbor search is employed to retrieve the most similar documents.

3.7   Sentence as Vector

We tried two approaches to obtain sentence embeddings.

Firstly, if the length of the sentence is greater or equal than 3, we average sentence terms' embeddings to obtain sentence embeddings. Before that, words with low IDF (<0.01) are removed from sentences.

Secondly, we use pretrained sentence representation model - LASER [3]. LASER is sentence encoder supporting 93 languages trained on parallel sentences. One detail that should be mentioned is that we used this model on normalized sentences, whereas it was trained on unnormalized ones. It could decrease quality of sentence embeddings. Also dimensionality of embeddings from LASER is larger than our embeddings (1024 vs 384).

Those vectors also indexed with ANN index. At search time, we search for the m most similar sentence vectors for each query sentence. The score of each document is calculated by the following formula:

$$rel(d) = \sum_{s \in F_d} sim_s \tag{2}$$

where d is a found document, F_d is a set of sentences that were found for the document d, sim_s is a similarity score between a found sentence and the query sentence. Since F_d is a set, it is enforced that each sentence from the found document is matched only with one sentence from the query document.

3.8   Explicit Semantic Analysis (ESA)

We implemented CL-ESA method described in [16] and firstly introduced for solving monolingual semantic relatedness task in [10]. This method represents the document as a weighted vector of concepts. Concepts are defined by Wikipedia articles. In the original work, the authors used all English Wikipedia articles as concepts. We selected around 800 000 English articles that are aligned with Russian Wikipedia articles (articles that identified as comparable across languages by the Wikipedia community). For a given document D the weight of a concept C is defined as cosine similarity between top M keywords of D and matched keywords of an article that is linked with the concept C:

$$\frac{\sum_{w_i \in D} v_i \cdot c_i}{\sqrt{\sum_{w_i \in D} v_i^2} \sqrt{\sum_{w_i \in D} c_i^2}} \tag{3}$$

where v_i is the weight of a word w_i for D (e.g. TF-IDF), c_i is the weight of a word w_i for a Wikipedia article linked with the concept (e.g. TF-IDF).

We precomputed vector of concepts for each document in text collection and stored them with the same inverted index implementation that was used for the inverted indexd based approach.

At query time, the query document is converted to a vector of weighted concepts, i.e., identificators of Wikipedia articles. Then those identificators are mapped to articles in the other language, and similar documents are retrieved via search in the inverted index.

3.9   Clustering of Word Embeddings

This approach is similar to ESA method, but instead of using concepts from Wikipedia, it uses centroids of clusterized embeddings. We used K-means method for clustering cross-lingual word embeddings into C clusters. We take top K keywords of the document and search n most similar centroids for each keyword. After that, we sum similarities to centroids to obtain a weighted vector of centroids for each document.

4   Datasets

4.1   Wiki Dataset

As in our previous work, we use Russian-English aligned Wikipedia articles as a dataset for evaluation of retrieval methods (Wikipedia dump of March 2020). We exclude all articles, which title starts with words "List of", which size in symbols is less than 800, and which number of sentences is less than 10. Then we divide all remaining pairs of articles into two groups and each group into five bins by the size of a Russian article in sentences. Statistic of the dataset is provided in the Table 3 and Table 4.

- comparable by size: those articles that satisfy the following requirement:

$$|len(a_{\mathrm{ru}}) - len(a_{\mathrm{en}})| < min(len(a_{\mathrm{ru}}), len(a_{\mathrm{en}}))/4 \qquad (4)$$

- non-comparable by size: Those articles that satisfy the following requirement:

$$|len(a_{\mathrm{ru}}) - len(a_{\mathrm{en}})| > min(len(a_{\mathrm{ru}}), len(a_{\mathrm{en}})) \qquad (5)$$

Then we sampled 100 documents from each group. That gives us a dataset that contains 1000 document pairs[5].

---

[5] http://nlp.isa.ru/ru-en-src-retr-dataset/.

Table 3. Statistic of comparable by size articles

Size in ru sents	Count	Mean size of ru texts	Mean size of en texts
(9, 50]	62291	2560.34	2626.16
(50, 100]	20012	5878.33	5989.66
(100, 200]	9163	11100.2	11301.2
(200, 400]	3526	21275	21693.2
(400, 1000]	1628	43478.6	44023.1

Table 4. Statistic of non-comparable by size articles

Size in ru sents	Count	Mean size of ru texts	Mean size of en texts
(9, 50]	170902	2336.02	11471.6
(50, 100]	42958	6076.27	17771.8
(100, 200]	22491	11519.4	22309.7
(200, 400]	9318	21425.2	26847.4
(400, 1000]	3517	42744.7	26895

4.2 Essays Dataset

Essays dataset contains essays of different types. Some essays were manually written by students. Students had to search for sources in English and translate them. They were allowed to use translation services, but the adjustment of the translated text was required to produce correct Russian text. Some percent of sentences should have been translated without any automatic tools.

Also, we translated the text of sources from Paraplag dataset from Russian to the English language. Paraplag [21] is the monolingual dataset for evaluation of plagiarism detection methods. This dataset contains manually written essays on the given topic. Authors of essays should have used at least 5 sources, which they had to search by themselves when composing essays. The plagiarism cases vary by the level of complexity: from copy-paste plagiarism to heavily disguised plagiarism. We evaluate BLEU score of various translation services for Russian→English translation on a test dataset (Table 5). We used 300 sentences from WMT-News dataset[6], 300 sentences from News-Commentary dataset[7] and 300 sentences from cross-lingual essays.

In the end, we used Yandex Cloud translate API to translate 800 sources. It should be noted that we translated only sentences that were reused in essays and the nearest context of them (9500 characters above and below). The statistic of the translated dataset is presented in the Table 6.

---

[6] http://opus.nlpl.eu/WMT-News.php.
[7] http://opus.nlpl.eu/News-Commentary.php.

Table 5. BLEU scores of popular MT services.

Translation services	BLEU
Google translate	37.9
Yandex translate	38.61
Yandex cloud translate	40.32

Table 6. Statistic of essays dataset

	Essays			Sources		
	# of essays	Avg # of chars	Avg # of sents	# of sources	Avg # of chars	Avg # of sents
Copy-pasted/moderately disguised essays	50	18106	158	330	14000	108
Heavily disguised essays	74	14091	109	481	15737	123
Manually translated essays	26	19690	142	133	32544	367

5   Experiments Setup

5.1   Indexing of Wikipedia

We indexed all articles from English (6 M) and Russian (1.6 M) Wikipedia dumps (March 2020). In the preprocessing stage, we separate each list element with a double newline, to force the sentence termination for each list element. It avoids creating very long sentences that typically is produced from Wikipedia lists.

5.2   Document as Vector

We take top keywords with weight >0.04. We use Faiss GpuFlatIndex [13] for performing exhaustive search for similar document embeddings on GPU.

5.3   Sentence as Vector

All vectors were split into chunks with maximum size about 12.5 GiB. Each chunk was indexed independently using Faiss GpuIndexIVFPQ index with the following parameters: number of centroids - $5 * \sqrt{|V|}$, where V set of all vectors that we need to index; a training set contains all vectors; compression is PQ48 for our embeddings and PQ64 for LASER embeddings. Search parameters for each sentence: topk - 30, nprobe - 24.

5.4   ESA

When precomputing concept vectors for ESA method, we used 200 top keywords (with weight >0.05) of a document to compute weights of concepts. We kept the maximum 1200 concepts with the largest weight per document. Since we build vectors of Wikipedia articles using Wikipedia articles as concepts, we excluded a concept that represents the same article from the vectors.

5.5   Clustering of Word Embeddings

We chose 40k clusters for one million word vectors. We used 250 top keywords (with weight >0.04) to calculate a vector of weighted centroids. We kept only 100 centroids with the largest weight for each document.

5.6   Parameters Tuning

We used a grid search for parameters tuning on 400 wiki documents that were sampled independently of the testing data.

6   Evaluation Results

We performed a search on all documents from two datasets using various methods. All sources from two datasets were combined into one collection. The most similar 150 documents were retrieved and evaluated by standard metrics: Recall, MAP. We use the following abbreviation in the following table and below:

- IIM - Inverted Index based method.
- IIM-m - Translation method
- MT - Machine Translation
- DocVec - document as vector
- SentVec - sentence as vector
- ESA - explicit semantic analysis
- Cent - clustering method

We also measured the search speed of all methods on the wiki dataset (Time column). The were 10 parallel threads while searching in an inverted index and vectors were process in batches of size 10. Tests were run on common hardware: Core i9-9900K, 64 GiB RAM, 2 GPU GeForce RTX 2080.

Table 7 displays the evaluation results obtained on the wiki and essays datasets.

The results show that the translation method with BM25 ranking function is better in terms of Recall and Map than other methods on both datasets. The IIM with BM25 is slightly worse than IIM-m. Other similarity functions give worse performance, especially in terms of MAP. A simple method based on machine translation can't compete with the methods based on embeddings. Also, the machine translation of a query document is very computationally intensive and therefore, it is much slower than embeddings based methods (althoug MT was run on GPU). Vectors obtained via LASER shows best performance among vector-based methods. The performance of vector based methods is ruined when

Table 7. Evaluation results.

Method	Essays				Wiki				Time (s)
	P@1	Rec@10	Rec	MAP	P@1	Rec@10	Rec	MAP	
IIM-m, bm25	**0.986**	**0.921**	**0.994**	**0.912**	**0.725**	**0.871**	**0.924**	**0.777**	111
IIM, bm25	0.986	0.921	0.994	0.912	0.693	0.864	0.924	0.753	62.95
MT, bm25	0.846	0.856	0.96	0.804	0.611	0.779	0.889	0.67	350
IIM-m, cos	0.713	0.613	0.859	0.52	0.578	0.758	0.874	0.64	26.4
IIM, cos	0.726	0.616	0.874	0.524	0.595	0.782	0.878	0.66	58.5
MT, cos	0.54	0.425	0.759	0.344	0.394	0.592	0.789	0.461	372
IIM-m, ham	0.646	0.553	0.832	0.453	0.567	0.767	0.871	0.634	24.8
IIM, ham	0.666	0.583	0.837	0.747	0.575	0.763	0.89	0.642	46.79
MT, ham	0.606	0.389	0.681	0.308	0.401	0.575	0.742	0.461	358
DocVec (GPU)	0.38	0.259	0.534	0.196	0.307	0.495	0.713	0.375	3.8
SentVec (GPU)	0.28	0.335	0.648	0.241	0.246	0.491	0.685	0.33	43.34
LASER (GPU)	0.713	0.672	0.8	0.587	0.432	0.615	0.719	0.495	51.6
Cent	0.44	0.271	0.435	0.217	0.161	0.306	0.479	0.212	59.64
ESA	0.526	0.407	0.771	0.324	0.262	0.465	0.701	0.331	23.14

Table 8. Performance for essays dataset only

	Rec	MAP
DocVec	0.99	0.81
SentVec	0.99	0.89
Cent	0.99	0.74

searching over a large number of source vectors. For example, when searching only on sources from the essays dataset, the performance of those methods is much better (Table 8).

ESA shows good recall, but ranking of found documents is worse than for the IIM and vector based methods.

It should be pointed out that the performance of the methods differs depending on the size of the documents (Table 9).

Methods work better with long texts that are comparable by size. Short texts (groups 1,2) are likely to have some specific out-of-vocabulary lexis, and remaining words do not help to retrieve the article in other language and rank it highly. Non-comparable by size text has lower MAP than comparable ones for all size bins. One of the lowest MAP is for the combination 9, i.e. the long texts, whereas the best MAP is for the longest texts also, but this time for comparable ones (group 10). It can be seen from Table 4 that Russian texts from the non-comparable group are longer than English texts by the factor of two. These articles may be devoted to Russian concepts that have short descriptions in English. In this case, similar articles with longer texts have more chances to share lexis than shorter articles. Therefore, shorter texts are lower in the rank list. Same behavior is observed for other methods too. For example, Table 10 presents the comparison of results for SentVec method.

Table 9. Metrics per each size group (IIM, bm25 method)

No	Size in ru sents	Comparable by size?	MAP	Rec
1	(9, 50]	False	0.792	0.939
2	(9, 50]	True	0.756	0.89
3	(50, 100]	False	0.679	0.92
4	(50, 100]	True	0.796	0.95
5	(100, 200]	False	0.664	0.89
6	(100, 200]	True	0.869	0.96
7	(200, 400]	False	0.673	0.9
8	(200, 400]	True	0.879	0.969
9	(400, 1000]	False	0.648	0.888
10	(400, 1000]	True	0.890	0.989

Table 10. SentVec metrics per each size group

No	Size in ru sents	Comparable by size?	MAP	Rec
1	(9, 50]	False	0.181	0.46
2	(9, 50]	True	0.345	0.53
3	(50, 100]	False	0.245	0.6
4	(50, 100]	True	0.360	0.63
5	(100, 200]	False	0.297	0.66
6	(100, 200]	True	0.515	0.78
7	(200, 400]	False	0.312	0.65
8	(200, 400]	True	0.528	0.9
9	(400, 1000]	False	0.162	0.67
10	(400, 1000]	True	0.684	0.93

The gap in performance between non-comparable and comparable texts is even larger for this method.

7   Conclusion

In this article, we compared various methods for cross-lingual retrieval of similar documents. We employed classical inverted indexes combined with cross-lingual embeddings and pure continuous retrieval using ANN. For this task and our datasets, the best result was shown by inverted index based approach. Dealing with OOV is another important issue. One way to solve it is to employ subword vector representation to encode the OOV-words [6]. Another way is to extending vocabulary from different comparable corpora: scientific papers, patents, etc. containing a lot of special lexis and terms. One of the possible solutions is to use the system for translated plagiarism detection to extract parallel sentences from comparable corpora [19, 30].

Acknowledgement. The reported study was funded by RFBR according to the research projects No 18–37-20017 & No 18–29-03187. This research is also partially supported by the Ministry of Science and Higher Education of the Russian Federation according to the agreement between the Lomonosov Moscow State University and the Foundation of project support of the National Technology Initiative No 13/1251/2018 dated 11.12.2018 within the Research Program "Center of Big Data Storage and Analysis" of the National Technology Initiative Competence Center (project "Text mining tools for big data").

References

1. Antonova, A., Misyurev, A.: Building a web-based parallel corpus and filtering out machine-translated text. In: Proceedings of the 4th Workshop on Building and Using Comparable Corpora: Comparable Corpora and the Web, pp. 136–144 (2011)
2. Artetxe, M., Labaka, G., Agirre, E.: Generalizing and improving bilingual word embedding mappings with a multi-step framework of linear transformations. In: AAAI, pp. 5012–5019 (2018)
3. Artetxe, M., Schwenk, H.: Massively multilingual sentence embeddings for zero-shot cross-lingual transfer and beyond. Trans. Assoc. Comput. Linguist. **7**, 597–610 (2019)
4. Bakhteev, O., Ogaltsov, A., Khazov, A., Safin, K., Kuznetsova, R.: Crosslang: the system of cross-lingual plagiarism detection. In: Workshop on Document Intelligence at NeurIPS 2019 (2019)
5. Barrón-Cedeño, A., Gupta, P., Rosso, P.: Methods for cross-language plagiarism detection. Knowl.-Based Syst. **50**, 211–217 (2013)
6. Bojanowski, P., Grave, E., Joulin, A., Mikolov, T.: Enriching word vectors with subword information. Trans. Assoc. Comput. Linguist. **5**, 135–146 (2017)
7. Conneau, A., Lample, G., Ranzato, M., Denoyer, L., Jégou, H.: Word translation without parallel data. arXiv preprint arXiv:1710.04087 (2017)
8. Ferrero, J., Agnes, F., Besacier, L., Schwab, D.: Usingword embedding for cross-language plagiarism detection. arXiv preprint arXiv:1702.03082 (2017)
9. Franco-Salvador, M., Gupta, P., Rosso, P., Banchs, R.E.: Cross-language plagiarism detection over continuous-space-and knowledge graph-based representations of language. Knowl.-based Syst. **111**, 87–99 (2016)
10. Gabrilovich, E., Markovitch, S., et al.: Computing semantic relatedness using wikipedia-based explicit semantic analysis. IJcAI. **7**, 1606–1611 (2007)
11. Gillick, D., Presta, A., Tomar, G.S.: End-to-end retrieval in continuous space. arXiv preprint arXiv:1811.08008 (2018)
12. Jiang, J.Y., Zhang, M., Li, C., Bendersky, M., Golbandi, N., Najork, M.: Semantic text matching for long-form documents. In: The World Wide Web Conference, pp. 795–806 (2019)
13. Johnson, J., Douze, M., Jégou, H.: Billion-scale similarity search with gpus. IEEE Trans. Big Data (2019)
14. Kutuzov, A., Kopotev, M., Sviridenko, T., Ivanova, L.: Clustering comparable corpora of russian and ukrainian academic texts: Word embeddings and semantic fingerprints. arXiv preprint arXiv:1604.05372 (2016)

15. Mikolov, T., Sutskever, I., Chen, K., Corrado, G.S., Dean, J.: Distributed representations of words and phrases and their compositionality. In: Advances in Neural Information Processing Systems, pp. 3111–3119 (2013)
16. Potthast, M., Barrón-Cedeño, A., Stein, B., Rosso, P.: Cross-language plagiarism detection. Lang. Res. Eval. **45**(1), 45–62 (2011)
17. Rekabsaz, N., Lupu, M., Hanbury, A., Zuccon, G.: Generalizing translation models in the probabilistic relevance framework. In: Proceedings of the 25th ACM International on Conference on Information and Knowledge Management, pp. 711–720 (2016)
18. Romanov, A., Kuznetsova, R., Bakhteev, O., Khritankov, A.: Machine-translated text detection in a collection of russian scientific papers. Dialogue, p. 2 (2016)
19. Schwenk, H., Chaudhary, V., Sun, S., Gong, H., Guzmán, F.: Wikimatrix: Mining 135m parallel sentences in 1620 language pairs from wikipedia. arXiv preprint arXiv:1907.05791 (2019)
20. Sochenkov, I.V., Zubarev, D.V., Tikhomirov, I.A.: Exploratory patent search. Inf. its Appl. **12**(1), 89–94 (2018)
21. Sochenkov, I., Zubarev, D., Smirnov, I.: The paraplag: Russian dataset for paraphrased plagiarism detection. In: Computational Linguistics and Intellectual Technologies: Papers from the Annual International Conference Dialogue, vol. 1, pp. 284–297 (2017)
22. Straka, M., Hajic, J., Straková, J.: Udpipe: trainable pipeline for processing conll-u files performing tokenization, morphological analysis, pos tagging and parsing. In: Proceedings of the Tenth International Conference on Language Resources and Evaluation (LREC 2016), pp. 4290–4297 (2016)
23. Tang, S., Mousavi, M., de Sa, V.R.: An empirical study on post-processing methods for word embeddings. arXiv preprint arXiv:1905.10971 (2019)
24. Tiedemann, J.: Parallel data, tools and interfaces in opus. Lrec **2012**, 2214–2218 (2012)
25. Vulić, I., et al.: Multi-simlex: A large-scale evaluation of multilingual and cross-lingual lexical semantic similarity. arXiv preprint arXiv:2003.04866 (2020)
26. Vulic, I., Moens, M.F.: Bilingual word embeddings from non-parallel document-aligned data applied to bilingual lexicon induction. In: Proceedings of the 53rd Annual Meeting of the Association for Computational Linguistics (ACL 2015), vol. 2, pp. 719–725. ACL; East Stroudsburg, PA (2015)
27. Vulić, I., Moens, M.F.: Monolingual and cross-lingual information retrieval models based on (bilingual) word embeddings. In: Proceedings of the 38th International ACM SIGIR Conference on Research and Development in Information Retrieval, pp. 363–372 (2015)
28. Zubarev, D.V., Sochenkov, I.V.: Cross-lingual similar document retrieval methods. In: Proceedings of the ISP RAS, 31(5) (2019)
29. Zubarev, D., Sochenkov, I.: Cross-language text alignment for plagiarism detection based on contextual and context-free models. In: Proceedings of the Annual International Conference Dialogue, vol. 1, pp. 799–810 (2019)
30. Zweigenbaum, P., Sharoff, S., Rapp, R.: Overview of the third bucc shared task: spotting parallel sentences in comparable corpora. In: Proceedings of 11th Workshop on Building and Using Comparable Corpora, pp. 39–42 (2018)

Author Index

Printed in the United States
by Baker & Taylor Publisher Services